Concepts and development of highly integrated payloads for space missions on the example of the microrover NANOKHOD

A thesis accepted by the Faculty of Aerospace Engineering and Geodesy of the Universität Stuttgart in partial fulfilment of the requirements for the degree of Doctor of Engineering Sciences (Dr.-Ing.)

by

Sabine Klinkner

born in Greenwich, CT.

Main referee: Prof. Dr. rer. nat. Hans–Peter Röser
Co–referee: Prof. Dr. Klaus Schilling
Date of defence: 11.12.2009

Institute of Space Systems
Universität Stuttgart
2009

Berichte aus der Luft- und Raumfahrttechnik

Sabine Klinkner

Concepts and development of highly integrated payloads for space missions on the example of the microrover NANOKHOD

D 93 (Diss. Universität Stuttgart)

Shaker Verlag
Aachen 2010

Bibliographic information published by the Deutsche Nationalbibliothek
The Deutsche Nationalbibliothek lists this publication in the Deutsche
Nationalbibliografie; detailed bibliographic data are available in the Internet at
http://dnb.d-nb.de.

Zugl.: Stuttgart, Univ., Diss., 2009

ISBN 978-3-8322-9557-8
ISSN 0945-2214

Shaker Verlag GmbH • P.O. BOX 101818 • D-52018 Aachen
Phone: 0049/2407/9596-0 • Telefax: 0049/2407/9596-9
Internet: www.shaker.de • e-mail: info@shaker.de

Contents

Contents

List of Figures

List of Figures

List of Tables

Abstract

Since the beginning of space exploration, the restriction on mass has been one of the strongest drivers. The need for increasing payload to resource mass ratios is still paramount for current extraterrestrial science and exploration missions. This becomes especially obvious for missions that are challenging when considering their power consumption. A mission to the planet Mercury, being the closest planet to the sun, is a good example for a mission with extremely high power needs, which are restricting the overall system mass.

Mercury's proximity to the sun imposes harsh environmental conditions – e. g. thermal issues combined with high vacuum – and challenging requirements for any mission to visit – e. g. gravitational potential of the sun. At the same time it makes the planet an interesting target for scientists. Mercury carries essential information on the evolution and origin of our solar system, which is again the key to understand how the conditions for life could develop. The realisation of a mission to explore Mercury is only possible by fulfilling difficult requirements and above all limiting the system mass.

Despite a limited mass, mobility is still a key feature, which is useful for any planetary science mission and for space exploration in general. Missions with mobile systems provide a much wider spectrum of outcomes by analysing a higher number of samples within an increased area of exploration. The additional degree of freedom of a microrover in comparison to a lander or even a robotic arm allows the mission scenario to be easily adapted to the landing site as it is encountered.

The ratio of scientific payload to resource mass for microrovers can be very efficient as demonstrated by the Nanokhod rover, built by the company *von Hoerner & Sulger GmbH* (vH&S), Schwetzingen, Germany. The microrover Nanokhod has been developed since the 1990's and adapted for several target surfaces like the Moon, Mars and now Mercury. The Nanokhod rover is a tethered system, which means it stays connected to the lander via two thin tether wires throughout the whole mission. This connection provides the rover to rely on synergy effects, by receiving its power from the lander and by transmitting all data via the lander's communication subsystems down to Earth. This results into a volume of only $160 \times 65 \times 250 \, mm^3$ with a peak power of 5 W for the microrover. The complete rover system weighs 3.2 kg, including a payload mass of 1 kg.

The newest development of the Nanokhod, described in this thesis, bases upon the boundary conditions of the ESA cornerstone mission BepiColombo. The Nanokhod was chosen as model payload for the surface element of the Mercury mission and was developed in the framework of the ESA TRP (Technical Research Project) project MRP (Mercury Robotic Payload) for the thus imposed mission requirements. Although the BepiColombo lander unfortunately was cancelled shortly after the MRP-project was initiated, the Mercury conditions were kept as the boundary conditions for the rover development. The BepiColombo mission scenario provides extremely difficult conditions, accordingly, a rover, which fulfils all these conditions, provides a good baseline design for other missions with comparatively easier requirements:

Due to the missing atmosphere, the landing procedure on the Mercury surface cannot use break parachutes, but has to rely on chemical propulsion only. To still minimise the amount of propulsion for the landing, the mission requires a minimum system mass and volume. For a further reduction the mission cannot consider a soft landing scenario; instead, all landing elements have to be robust enough to survive a landing shock of 200 g for 20 ms.

The power consumption of the rover system also has to be reduced to a minimum. The mission is targeted for the night side of the planet and thus all lander subsystems rely on the power supply of primary batteries and thus the energy consumption has a direct impact on the overall battery mass. With the Nanokhod receiving all its energy from the lander battery, the electrical rover system has to be redesigned, in order to minimise the power needs of the rover.

The surface temperature of approx. $-180°C$ on the night side of Mercury is very demanding for the rover design. The negligible atmosphere worsens the thermal conditions additionally, because all heat dissipation of electronic parts can only be removed via thermal radiation and heat conduction through the rover structure. This limitation of heat transfer imposes the risk of hot areas within the rover, despite the extremely low surface temperatures. Therefore special attention has to be given to design the heat conductance interfaces effective enough.

It is not possible to implement an active thermal control system in this small highly integrated system, due to the limited volume, mass and power. Accordingly the rover system has to be based on a mainly passive thermal control system and all components have to be designed to work reliably for the complete temperature range starting from the minimum temperature of $-180°C$.

These extremely difficult boundary conditions require a special design philosophy for the development of the rover. The Mercury mission imposes a considerable amount of very ambitious mission requirements. Despite the fact that the main mission objective is scientific research, most of those requirements are determined by the harsh environmental conditions. For various critical subsystems the finding of a feasible solution becomes a characteristic feature of the design philosophy. In order to meet all these requirements, the rover design of former Nanokhod development projects was completely revised within the framework of the MRP-project.

In the course of this latest development the rover gained a near flight readiness with a practical design to withstand the tough requirements of a flight model. Having realised a design that meets the flight–model requirements, a hardware model has been manufactured to an engineering level, which is suitable for environmental testing of vibration, shock and thermal vacuum.

This thesis describes this latest rover design, the manufactured model and the lessons learnt from the system for the required environmental conditions. It also includes the development of a detailed thermal mathematical model of the rover system and the initial thermal vacuum test with a moderated thermal range. The initial thermal model development complemented with the output of the thermal vacuum test resulted in a reliable and detailed thermal model of the rover system, which is easily adaptable to any mission profile.

Abstract

Zusammenfassung

Masse ist in der Raumfahrt weiterhin einer der kritischsten Faktoren, den es auch heute immer weiter zu reduzieren gilt. Aus diesem Grund ist die Miniaturisierung in allen Teilbereichen der Raumfahrt ein Arbeitsfeld, das stetig vorangetrieben wird. Besonders für energetisch anspruchsvolle Missionen muss das Verhältnis von Nutzlast– zu Systemmasse optimiert werden, da die Gesamtmasse stark begrenzt ist. Ein Beispiel für eine solche Mission, deren hoher Energiebedarf die Systemmasse limitiert, ist eine Mission zum Planeten Merkur.

Die Nähe des Planeten zur Sonne verursacht schwierigste Umweltbedingungen auf der Oberfläche des Merkur (z.B. extreme Temperaturschwankungen in Kombination mit Vakuum) und hohe Anforderungen für jede Mission mit diesem Planeten als Ziel (z.B. der energetisch aufwändige Transfer, um das hohen Gravitationspotenzial zwischen Erde und Merkur zu überwinden). Gleichzeitig ist der Planet aber ein hochinteressantes wissenschaftliches Forschungsgebiet, da er grundlegende Informationen über die Entwicklung und den Ursprung unseres Sonnensystems birgt. Diese Informationen sind der Schlüssel zum Verständnis, wie sich die Bedingungen für die Entstehung von Leben entwickeln konnten. Um jedoch eine solch wissenschaftlich interessante Mission zu ermöglichen, muss die Masse aller Missionselemente minimiert werden.

Trotz einer stark begrenzten Gesamtmasse, die sich für die Realisierung einer Merkurmission ergibt, ist Mobilität ein weiteres Hauptmerkmal für Wissenschafts– und Explorationsmissionen. Missionen, die über mobile Systeme verfügen, bieten ein deutlich breiteres Ergebnisspektrum durch die Untersuchung von einer höheren Anzahl von Proben in einem erweiterten Forschungsbereich. Der zusätzliche Freiheitsgrad eines Microrovers im Vergleich zu einem Lander oder einem robotischem Arm bietet dem System die Möglichkeit, sich flexibel an die vorgefundenen Bedingungen des Landeplatzes anzupassen.

Weiterhin kann bei einem Microrover das Verhältnis von wissenschaftlicher Nutzlast zu Gesamtmasse sehr effizient gestaltet werden, wie es der Nanokhod–Rover der Firma *von Hoerner & Sulger GmbH* (vH&S), Schwetzingen, Deutschland, eindrucksvoll demonstriert. Der Nanokhod wird seit den 1990er Jahren bei vH&S entwickelt und ist in der Zeit für die verschiedensten Oberflächen von z. B. Mond, Mars und nun Merkur ausgelegt worden. Der Microrover Nanokhod ist ein

Tether–System, das während der gesamten Mission über zwei dünne Tetherdrähte mit dem Lander verbunden bleibt. Die Verbindung mit dem Lander ermöglicht dem Roversystem Synergieeffekte zu nutzen, indem er für die Datenübertragung den Lander für die Übermittlung benutzt und für die Stromversorgung die Energiesubsysteme des Landers verwendet. So ergibt sich ein Volumen von nur $160 \times 65 \times 250 \, mm^3$ für den Microrover, bei einem Spitzenverbrauch von 5 W. Das Systemgewicht des Fahrzeugs ist 3.2 kg und beinhaltet eine Nutzlastmasse von 1 kg.

Die in dieser Arbeit beschriebene Neuentwicklung des Nanokhods basiert auf den Rahmenbedingungen der europäischen Merkur–Mission BepiColombo. Der Nanokhod war als Modellnutzlast für das Landeelement der Mission ausgewählt und wurde im Rahmen des ESA TRP[1]-Projektes MRP (Mercury Robotic Payload) für die Anforderungen des Merkur entwickelt. Obwohl der BepiColombo-Lander kurz nach Beginn des MRP-Projektes abgesagt worden ist, wurden die Merkurbedingungen für die Auslegung des Rovers beibehalten. Das BepiColombo Missionsszenario stellt außergewöhnlich schwierige Anforderungen an das Roversystem, dementsprechend bietet ein Rover, der diese Bedingungen erfüllt, eine hervorragende Grundlage für Missionsziele mit vergleichsweise verminderten Anforderungen:

Durch die fehlende Atmosphäre kann der Landeanflug auf den Merkur nicht durch Bremsfallschirme verzögert werden, sondern muss ausschließlich mit chemischen Antrieben abgebremst werden. Um den Treibstoffbedarf für die Landung zu minimieren, wird, neben der Forderung nach minimaler Systemmasse und kleinstmöglichen Volumen, keine weiche Landung angestrebt. Stattdessen müssen alle Landelemente robust genug gebaut werden, um einen Landeschock von 200 g für 20 ms zu überstehen.

Im Zusammenhang mit der möglichst geringen Systemmasse steht die Anforderung, den Energiebedarf der Rovers soweit wie möglich zu reduzieren. Da die Mission ausschließlich für die Nachtseite des Planeten geplant ist, sind alle Lander Subsysteme auf die Versorgung aus Primärbatterien angewiesen. Auch der Nanokhod bezieht alle notwendige Energie direkt aus der Landerbatterie. Der Energiebedarf des Rovers steht somit in direktem Bezug zu der Masse der Landerbatterie, und muss folglich für die gesamte Missionsdauer minimiert werden.

Die Oberflächentemperatur von ca. $-180\,°C$ auf der Nachtseite des Merkur stellt hohe Anforderungen an das Roverdesign. Die Vakuumbedingungen des Merkur verschlechtern die thermischen Bedingungen im Roversystem zusätzlich,

---

[1]Technical Research Project – technisches Forschungsprojekt

da die Wärmedissipation der elektronischen Bauteile ausschließlich durch Wärmestrahlung oder Wärmeleitung innerhalb der Roverstruktur abgeführt werden kann. Das bedeutet, dass, trotz der extrem tiefen Temperaturen, Teile des Rovers zu heiß werden können, wenn die Wärmeabfuhr nicht effektiv genug gestaltet ist.

Aufgrund der Forderung nach geringer Masse ist es nicht möglich ein aktives Thermalsystem in diesem hochintegriertem Mikrosystem unterzubringen. Dementsprechend müssen, bei der Nutzung eines weitgehend passiven Thermalkontrollsystems, alle Roverkomponenten so ausgelegt werden, dass sie von der Minimaltemperatur von $-180\,^\circ$C an aufwärts funktionieren.

Diese äußerst schwierigen Rahmenbedingungen erfordern eine eigene Designphilosophie für die Entwicklung des Rovers. Der Anforderungskatalog für den Rover bei dieser Mission ist sehr umfangreich und gleichzeitig extrem anspruchsvoll. Obwohl die Mission ausschließlich der Erforschung des Merkur dient, ist nur der kleinere Teil der Anforderungen durch die wissenschaftlichen Ziele definiert; in weiten Bereichen ist die Entwicklung des Rovers von den komplexen Umweltbedingungen der Merkurmission bestimmt. Das Finden einer technisch machbaren Lösung wird zum Hauptmerkmal der Designphilosophie. Um all diese Anforderungen zu erfüllen, wurde das bestehende Roverdesign aus vorangegangenen Projekten im Rahmen des MRP–Projektes vollständig überarbeitet.

Im Laufe dieses letzten Entwicklungsschrittes hat der Rover annähernd Flugreife erreicht. Ein praktikables Design ist entwickelt worden, das die schwierigen Anforderungen eines Flugmodells erfüllt. Basierend auf dem flugfähigen Design wurde ein Ingenieurmodell gebaut, das für die Durchführung von Vibrations–, Schock– und Thermal–Vakuum–Tests geeignet ist.

In dieser Arbeit wird das neueste Design des Rover Modells für die Anforderungen der Merkuroberfläche und dessen technische Realisierung, sowie die getroffenen Vorbereitungen für die anstehenden Umwelttests beschrieben. Ebenfalls Teil dieser Arbeit ist die Entwicklung eines detaillierten Thermalmodells des Rovers und die Durchführung eines Thermal–Vakuum–Umwelttests mit leicht reduzierten Anforderungen. Das ursprüngliche Thermalmodell, ergänzt durch die Ergebnisse aus dem Thermaltest, ermöglicht die Entwicklung eines detaillierten und verlässlichen Knotenmodells des Roversystems, welches leicht an jedes Missionsprofil angepasst werden kann.

Acronyms

ADC	Analogue–to–Digital Converter
AIV	Assembly Integration and Verification
AMS	Aerospace Material Standard (Issued by SAE)
APXS	Alpha–Particle–X–Ray Spectrometer
ASIC	Application Specific Integrated Circuit
AU	Astronomical Unit
B.C.	Before Christ
BIP-type	Bipolar type
BLDC	Brushless Direct Current Motor
CA	Cable Arm
CAD	Computer Aided Design
CDR	Critical Design Review
CNES	Centre National d'Etudes Spatiales (French Space Agency)
CoM	Centre of Mass
CPM	Chemical Propulsion Module
CS	Circular Spline (part of Harmonic Drive)
CTE	Coefficients of thermal expansion
Cu–LS	Copper silk woven strand
CuBe	Copper Beryllium
DARA	Deutschen Agentur für Raumfahrtangelegenheiten (former German Space Agency)
DC	Direct Current
DFKI	Deutsches Forschungsinstitut für künstliche Intelligenz, Bremen, Germany
DLR	Deutsches Zentrum für Luft– und Raumfahrt (German Aerospace Center)
ECSS	European Cooperation of Space Standardisation
EDLS	Entry, Descent and Landing System
EGSE	Electrical Ground Support Equipment
EMC	Electromagnetic Compatibility
EMF	Electro–Magnetic Force
EOL	End of Life
EPFL	Ecole Polytechnique Fédérale de Lausanne

EQM	Engineering Qualification Model
ESA	European Space Agency
ESD	Electrostatic Discharge
ETHZ	Eidgenössische Technische Hochschule Zürich
FFT	Fast Fourier Transformation
FS	Flexspline (part of Harmonic Drive)
GaAs	Gallium arsenide, semiconductor material
GIPF	Geochemistry Instrument Package Facility – ESA TRP study
GSE	Ground Support Equipment
GUI	Graphical User Interface
HD	Harmonic Drive
HDAG	Harmonic Drive AG
HDD	Hold–Down–Device (to secure the rover inside the lander)
HF	High Frequency
HFUC	Harmonic Drive component set series
HICoPS	Highly Integrated Communalised Payload System – ESA TRP study
I/F	Interface
IC	Integrated Circuit
I^2C	Inter–Integrated Circuit, type of serial communication protocol
IDD	Instrument Deployment Device
IRS	Institute of space systems (Universität Stuttgart)
ISAS	Institute of Space and Astronautical Science (part of the Japanese Space Agency JAXA)
ISD	Institute of statics and dynamics (Universität Stuttgart)
JAXA	Japan Aerospace Exploration Agency (Japanese Space Agency)
JPL	Jet Propulsion Laboratory
LAAS–CNRS	Laboratoire d'Analyse et d'Architecture des Systemes – Centre National de la Recherche Scientifique (French National Research Center, Analysis and System Architecture lab)
LLU	Locomotion Lever Unit (left Locomotion Unit with two motors)
LMS	Laser Mass Spectrometer
LOU	Locomotion Only Unit (right Locomotion Unit with only one motor)
LU	Locomotion Units

MCRT	Monte–Carlo ray tracing
MER	Mars Exploration Rover
Micro–RoSA	Micro–Robots for Scientific Applications – ESA TRP study
MIMOS	Mössbauer Spectrometer
MIROCAM	Microrover Camera
MIT	Massachusetts Institute of Technology
MLI	Multilayer Insulation
MM	Matrix Methods
MMO	Mercury Magnetospheric Orbiter (BepiColombo system element)
MOS	Metal Oxide Semiconductor
MPCh	Max–Planck–Institut für Chemie, Mainz, Germany
MPO	Mercury Planetary Orbiter (BepiColombo system element)
MRP	Mercury Robotic Payload – ESA TRP study
MSE	Mercury Surface Element (BepiColombo system element)
NASA	National Aeronautics and Space Administration – United States Space Agency
OLCS	On Lander Control Software
P/L	Payload
PA	Powered Arm
PCB	Printed Circuit Board
PEEK	Polyetheretherketone – a semicrystalline thermoplastic
PI	Principal Investigator
PLC	Payload Cabine
PMC	Planet Micro Cam – ESA TRP study
PSPE	Payload Support for Planetary Exploration – ESA TRP study
PTFE	Polytetrafluoroethylene – a synthetic fluoropolymer
RCL	Science & Technology Rover Co. ltd.
ROM	Rough Order of Magnitude
rpm	Revolutions per minute
RTG	Radioisotope thermoelectric generators
RTPE	Robotic Technology for Planetary Exploration – ESA TRP study
SAE	Society of American Engineers
SEPM	Solar Electrical Propulsion Module
SEU	Single event upsets
SiC	Silicon carbide
SME	Small and medium sized enterprise
SOI	Silicon on insulator

SSAC	Space Science Advisory Committee of ESA
TDU	Tether Data Unit (PCB for Tethercommunication)
TMM	Thermal Mathematical Model
TRP	Technology Research Programme of ESA
TU	Tether Unit
TV	Thermal Vacuum
VNII Transmash	Russian transport Machinery Engineering Institute
vH&S	von Hoerner & Sulger GmbH
WG	Wave Generator (part of Harmonic Drive)

1 Introduction

1.1 Motivation and background

Although Mercury is one of the planets already known by the ancient Greeks and has been observed since then by most ancient cultures and civilisations, it is still today relatively unexplored. Due to its vicinity to the sun, being its nearest neighbour, it is difficult to observe from Earth. From our planet it can only be seen in the twilight of the early morning and the late evening. Observations from the Earth orbit always bear the risk of damaging instruments by the intense solar radiation when pointing them towards the planet.

Mercury represents an extreme among the terrestrial planets, due to its composition and position in the solar system. It offers specific information on the formation and early history of our solar system and especially the inner planets. With the recent development in understanding the evolution of the solar system the interest in Mercury is again very high and several possible mission scenarios are currently foreseen to explore the innermost planet.

One of these missions is ESA's cornerstone mission BepiColombo. Initially this mission consisted of two orbiter elements and a surface element which were to carry out complementary analyses for the exploration of Mercury and the interplanetary cavity. As part of the BepiColombo surface element, the microrover Nanokhod of the company *von Hoerner & Sulger GmbH* (vH&S), Schwetzingen, Germany, was selected as model payload. The rover represents a mobile scientific platform for the exploration of the near surroundings of the lander. It was chosen as the rover provides an additional degree of freedom in comparison to a lander or even a robotic arm. It allows the mission to be easily adapted to the landing site as it is encountered. Missions with mobile systems provide a much wider spectrum of outcomes by exploring a higher number of samples within an increased area of exploration. Furthermore the ratio of scientific payload to resource mass for microrover can be very efficient as demonstrated by the Nanokhod.

A surface element which includes a rover also allows to obtain *ground truth* by direct measurements of the chemical and mineralogical composition as opposed to remote investigations from an orbiter. This is an essential scientific achievement [1]. The Nanokhod microrover with its geochemistry instrument package can address exactly this issue. The package consists of three miniaturised sensor heads which can analyse the chemical and geological characteristics. The basic philosophy of the rover system is thus not to collect samples and bring them to the lander but to transport a suite of instruments to different targets and perform in–situ measurements.

1.2 Objective and approach

This thesis describes the development of a mature rover for the Mercury environment based on the Mercury mission BepiColombo conditions and carries out a detailed thermal analyses for this system. Unfortunately, the Surface Element was cancelled from the BepiColombo mission in 2003 and with it this flight opportunity for the rover. Nevertheless, it was decided to continue the rover development for Mercury conditions in order to prepare it for future missions. This project helps solving detailed technical problems of a real implementation and thus gaining a better understanding for the system. A technology development for the challenging nature of the Mercury environment is also considered to be applicable with moderate modifications to a variety of other planetary bodies like Mars or Moon – targets with and without atmosphere. The developed rover has gained a near flight readiness with a practical design to withstand the harsh requirements of a flight model.

The framework conditions for a Mercury mission including a lander are very challenging. The transfer of an orbiter to Mercury is a very demanding task, which was so far only achieved by the American probe Mariner 10 and the current American mission Messenger. A mission to Mercury has to overcome the high gravitational potential of the sun from the Earth to Mercury. Finally, whilst in the Mercury orbit, the satellite would be exposed to high thermal variations between the night and day side of the planet. The Nanokhod described here is designed to endure both the journey conditions for up to 4 years as well as to land on its surface as part of the Mercury Surface Element.

The landing scenario evokes further significant requirements for the space system. As Mercury has no substantial atmosphere, the deceleration has to be performed with chemical propulsion. In order to reduce the amount of propulsion to a minimum the mission takes into account a high landing shock opposed to a soft landing scenario. After some initial studies of the rover mission under day as well as night side conditions of Mercury, the landing site was decided to be completely on the night side of the planet, where the absence of the sun's radiations reduces the thermal range and makes the mission feasible. This scenario becomes possible with the long sidereal Hermean[2] day lasting 88 Earth days. However the environment remains severe, coupling a high vacuum with surface temperatures estimated to $-180°$C.

All these requirements add up and generate two main design drivers for the surface element of a Mercury mission, which are mass and directly related to this, the volume. It becomes obvious that every kilogramme to land on the planet is very cost intensive, with an interplanetary journey from Earth to Mercury and the orbit entering phase requiring a high amount of fuel, together with a landing scenario which has to rely only on chemical means. Thus, mass and volume of the lander and rover have to be reduced to a minimum in order to realise such a mission. Opposing this, the rover has to be designed robust enough to cope with the vibration and shock environment of such a transfer and landing scenario.

Also related to the mass limitation issue is the energy consumption of the rover. The power is provided from the lander batteries and with a given mission duration, the consumed energy relates directly to the battery capacity and thus to the battery mass. The energy consumption is thus the next significant design driver of the rover system, requiring an electrical system, which is optimised for the specified mission profile.

This is also why the thermal control system of the rover has to be passive to keep the energy needs as low as possible. The passive system shall provide sufficient thermal conditions to allow the functioning of the rover without wasting limited energy resources. The whole system is designated to retain as much heat within the rover as possible. At the same time the design also avoids local hot spots, which can occur especially with active components in a vacuum environment. In addition to the extreme thermal environment the rover model is exposed to vacuum conditions on the Mercury surface. It has to withstand an expected landing shock of 200 g for 20 ms. On the surface it has to cope with very fine regolith which is abrasive and has an extremely low thermal conductivity.

Care was taken to design the critical components of the rover system for the harsh conditions. The mechanisms in particular are affected by the extremely low temperatures and the vacuum. In order to provide a reliable drive unit solution, a new drive combination was implemented and thoroughly tested.

Despite all these requirements imposed by the environmental and mission conditions, the main objective of the rover design is the accommodation and oper-

---

[2] *Hermean* = relating to Mercury

ation of the scientific instrument. However the design philosophy of this mission is driven by the considerable amount of ambitious mission requirements. Despite the scientific mission objective, most of those requirements are determined by the difficult environmental conditions. For various critical subsystems the finding of a feasible solution becomes a characteristic feature of the design philosophy.

Based on the design which is consistent with the described flight–model requirements, a hardware model has been manufactured to an engineering level which is suitable for environmental testing of vibration, shock and thermal vacuum. The design was done to allow for reasonable facilities to ease the assembly, integration and verification (AIV) of the flight model.

The rover is designed for a baseline mission of 14 Earth days with the possibility for a mission extension of another 14 days. The rover communicates and is powered by the lander via tether of at least 50 m which allows the rover to explore an area of at least 5 to 10 m radius from the landing point. The MRP Nanokhod shall be able to move across the Mercury surface with a velocity of 5 metres/hour and to negotiate steps of 10 cm in height and trenches of 10 cm in width. The mission foresees one in–situ analyses per Earth day with each of its three instruments: A microrover camera (MIROCAM), an Alpha–Particle X–ray Spectrometer (APXS) and a Mössbauer Spectrometer (MIMOS). The rover shall operate near–autonomously due to a single communication period once a day.

With extreme thermal environmental conditions, a preliminary thermal model was implemented during the design phase. This was necessary to have enough thermal information for the mechanical and electrical design of the rover. However, for the next development steps of the rover – the development and qualification of a full flight model using flight qualified electronic components – more knowledge of the temperature distribution is mandatory to avoid unnecessary iterations due to unforeseen thermal failures. Also the integration of the scientific payload into the rover system which has only a limited operating temperature range requires a detailed knowledge of the prevailing temperatures. All this information is needed as necessary input for the design of the thermal control concept of future rover development steps. Thus, a more sophisticated thermal model, which provides detailed information about the thermal behaviour of single components inside the Nanokhod, is necessary. A detailed thermal model also eases the adjustment of the rover design to other mission targets and thus makes the redesign of the system faster and less costly. For the first validation of the current rover design, as well as for the benefits of future developments, a detailed thermal model of the Nanokhod rover was developed within this work.

In order to have a reliable model, the development includes iterations that are based on the results gained from preliminary environmental tests in a thermal vacuum chamber. The thermal vacuum tests are at the same time a first qualification step of the rover model. Although the temperature range of the thermal vacuum test will not be fully representative of the Mercury conditions, this test will verify the electrical system and the rover mechanisms with the new drive units for low temperatures and vacuum. During the test the rover undergoes a functional test, which exercises the mechanism, the sensors and the electronics of the rover. During the functional test all parameters from the rover's housekeeping are logged to verify the functionality of all system components whilst in the test chamber.

1.3 Work content

The introduction in chapter 1 contains the background and frame conditions of the thesis on hand. It describes the origin of the Nanokhod rover development for the environmental conditions of the Mercury surface. It offers an overview of the rover development project, the mission it was designed for and on the objective and the approach of this thesis.

Chapter 2 considers the fundamentals of scientific robotic missions, which have to be taken into account for a rover development as described here. An overview on different robotic systems and robotic missions provides a background for the Nanokhod development and points out the surplus of a robotic element as part of science mission

A description of the microrover Nanokhod follows in chapter 3 which includes the development history of this rover system. Further the BepiColombo mission and the scientific relevance of such a mission are described in this chapter. On the basis of the development history of the Nanokhod and the BepiColombo mission the author compiles the requirements for the combined TRP projects MRP and GIPF. The requirements document is consistent to the Payload Definition Document of the BepiColombo mission.

The rover design for the Mercury mission was conducted in the frame work of the ESA TRP project MRP and is described in chapter 4. The project was carried out under the lead of the company *von Hoerner & Sulger GmbH*. Aside from three minor work packages which were accomplished by external institutes, the project was executed by a vH&S–team of 4 people. The author was responsible for the mechanical development of the rover system including the associated structural analyses. In a close cooperation with the company Harmonic Drive

AG, the author developed the new drive unit of the rover and integrated it into the system. The author also took care of the AIV and testing of the rover. In addition to the system test the author prepared a detailed analyses including a test programme for validation of the drive unit, which she conducted with the support of Harmonic Drive AG

Chapter 5 gives a general overview of the thermodynamic fundamentals of the thermal analyses. It describes the thermal control system of the rover and its background with the limitations of components and materials within the extreme environmental conditions. The chapter includes a description of the thermal rover design which the author developed and implemented.

The thermal model which was developed by the author in the framework of this thesis is depicted in chapter 6. The development of the initial thermal model was supported by a student research project, which was attended by the author. Different sections in the chapter describe the boundary conditions and applied assumptions of the model as well as the utilised software. The results of first material tests and model analyses, which were included into the thermal model development, were carried out by the author and are described in this chapter.

For the validation of the thermal model and for the verification of the thermal rover design, the author planned and conducted the first thermal vacuum tests with a slightly reduced range. The test and the description of the test conditions and the test setup are depicted in chapter 7. For the conduction of the tests at the test chamber of the IRS[3] the author was supported by the thermal–vacuum facility manager. Based on the gained test results the author performed two further iterations of the thermal model and discussed the comparison of the test and calculated results in chapter 7. The author iterated furthermore an optimised model by implementing the boundary conditions of the Mercury mission. The results of this model and the conclusions for the thermal rover design drawn from it are described in the last section of chapter 7.

The work conducted in the framework of the rover development is summarised in chapter 8. The concluding chapter also gives an outlook, in which the envisaged development steps towards a flight model are explained. Furthermore the chapter includes a description of alternative targets which can be anticipated for a Nanokhod mission which can be realised with only minor changes in the rover design.

---

[3]Institute of space systems, Universität Stuttgart

2 Space exploration with robotic systems

2.1 Definition of a space system

In order to define the Nanokhod rover, being the subject of this thesis, as a space system, it is necessary to define the phrasing *system* [2]. It is also necessary to define the influences and effects such a system has itself and has to be compliant to.

A system has one common goal, while consisting of several different components, defined as subsystems. Each subsystem has its own properties and functions. In between these subsystems there are a high number of interactions and exchanges. This means that the definition of a subsystem as well as the input and output values can change depending on the applied point of view.

The development of such a system has to be considered as a network type problem. The subsystems have to be defined and developed at the same time. This includes many different disciplines which influence the development. Such a development does not proceed sequentially but follows heuristic patterns towards the flight model.

The specific characteristics of space systems are the demanding requirements of the system's environments. A space system has to withstand difficult environmental conditions and at the same time is limited in resources. A full range of issues has to be considered due to the complexity of a space system. Mass reduction and miniaturisation have to be promoted in all disciplines which contribute to the design of a space system. However in addition to all these efforts in the individual disciplines, the main challenge is considering and dealing with the system as a whole and its internal interactions. This is particularly true for small space systems, which require greater use of highly integrated structures to achieve their goals.

2.2 Design strategies of highly integrated systems

For the development of a small space system, reduction in resource usage such as mass and volume, may be achieved through the optimisation if each individual sub–component of the system. However, all these efforts will only provide a certain level of reduction. A further reduction can be attained by using all possible synergies already at the lowest level of space system design. This means moving away from a design of discrete boxes. The need for highly integrated structures is especially evident for small space systems, as single items of the system take a much larger proportion of the system resources than in larger system. Thus a greater proportion of mass and volume can be saved by eliminating redundant functionality between components within a highly integrated system.

This approach of the system's design has extensive consequences in technical and organisational aspects. Payloads can no longer be chosen, designed and developed separately from the main system. The subsystems have to be analysed in the concept phase of the project. Main components, necessary for the scientific measurement task (the instrument cores), must be separated from the generic supporting components. While the instrument cores are developed individually and optimised for their specific measurement task mostly in a close cooperation with the responsible scientific institution, the generic instrument parts (e. g. for control, power and data management) are designed in one common subsystem with the target to maximise synergies and reduce the required resources to a minimum.

2.3 Robotic systems for space applications

With a history of space flight of more than 50 years, still space is a target which is difficult to reach and explore for mankind. The space environment is hostile and uncompromising for human life. In order to further explore our solar system, robotic systems assume a principle role in this task.

With the ongoing use of robotic systems within space flight, the systems themselves become more complex, but also notably more capable to take on more demanding tasks. Within space research areas there is a variety of different robotic systems handling all kinds of exploration's tasks and have various observation ranges, see also Figure 2.1.

2.3.1 Orbiter

The orbiter is the oldest application of robotic systems in space exploration, starting with the very first man–made space system, the Russian probe Sputnik 1. With an orbiter only remote sensing of the target celestial body is possible.

An orbiter offers good possibilities for repetitive measurements which allow a systematic approach of the exploration task. The orbiter provides a global perspective on the target planet. Due to reduced complexity, such a system is able to work for a long time in the orbit, however depending on the gravitational field of the lunar/planetary body.

2.3.2 Balloon

For the exploration of target planets with an atmosphere several concepts using a balloon have been developed. Balloons offer the possibility to get from a regional up to global view of the target planet, depending on the system implemented. Again the measurements can be carried out in a repetitive and thus very systematic way. However, the trajectory control is only limited, as balloons are dependent on the influences of weather conditions. Some balloon systems even allow for a limited amount of in–situ surface operations. Balloon systems have been frequently used for exploration within the earth atmosphere.

2.3.3 Aeroplane

There have been plans to carry out a planetary exploration of Mars with the help of aeroplane robots. The advantage is that no landing system is necessary; the system still has to ingress the atmosphere, but it needs no additional equipment for landing.

For the planetary surface, only remote sensing is possible with an aeroplane system, as such a system does not have any ground access, unless at the end of mission. Such a system is predestined for in–situ atmospheric exploration, but also magnetic, gravitational and electric sensing can be carried out with such a system. An aeroplane system has typically a reach of 100 to 1000 km. The exploration of the planetary body of interest with a plane can be done in a repetitive and systematic approach.

2.3.4 Helicopter

Helicopter robots have been planned for the exploration of the Martian surface and atmosphere. These systems have the same advantage as the aeroplanes: They do not need a complex airbag system for the landing. However, a helicopter is a fairly complex system to carry out analyses on an autonomous base in an unknown environment.

Planetary helicopters have a reach of 10 to 100 km though with the possibility of multiple ground access. This allows for a repetitive systematic in–situ analyses.

2.3.5 Penetrator

A penetrator is a space probe which is a projectile equipped with scientific instruments. Generally a high number of penetrators, which are all equipped with the same instrumentation, are used as part of the same mission and distributed over a region or even the whole surface of the target planet. The penetrators are designed to penetrate into the surface, rather than to land on the surface. Once within the surface they carry out their measurements.

With penetrators, the access to the target can be regional up to global, depending on the number of systems and the method of deployment used. Like that it is also possible to access multiple sample sites, making the system not a systematic, but stochastic way of surface exploration.

2.3.6 Lander

A lander has the possibility to access only one single surface location, which furthermore can only be chosen to the precision of a landing ellipse. The landing ellipse results from variable variations during the landing, which cause an *off–target* movement. Some possible variables are e. g. the entry angle, the entry mass, the atmosphere and the drag. The necessary landing ellipse may impose stringent requirements of the landing environment which may limit the number of possible regional areas. This requires a careful choice of the landing point from only a limited number of regional areas. A simple lander thus only offers very local information of the target planet. The reach of such a system can be slightly augmented with the integration of a robotic arm, but it still stays limited to only a small radius.

2.3.7 Rover

A rover for space exploration is any vehicle moving across the surface of a planetary body to carry out investigations. Depending on the rover system used the exploration abilities reach from local up to regional mobility. The starting point will always be a lander, so again the area where the rover starts its mission will have the same restriction as for the lander, such as the landing ellipse. Some systems with regional range can even compensate the uncertainty of the landing ellipse, depending on the systems terrainability and on the encountered environmental conditions. The exploration carried out by a rover is then repetitive and systematic within the local, respectively the regional range.

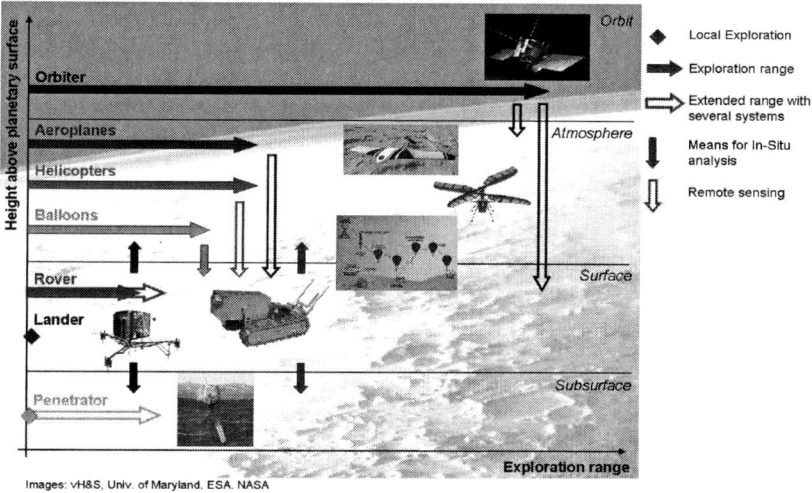

Figure 2.1: Exploration range of the different robotic systems

2.4 Rovers as space systems

In addition to the robotic systems mentioned in section 2.3, there is a whole variety of space systems belonging to the category of rover systems.

Rover systems are not limited to rolling systems (wheeled or tracked rovers), but also include concepts of legged systems. Apart from these most commonly considered principles, there are also jumping and crawling systems and a variety of hybrid chassis. These later systems are often discarded due to their complexity. All of these rover systems though offer different locomotion capabilities and features like [3]:

- Amount of articulators.
- Drawbar pull.

- Terrainability.
- Power supply and needs.
- Data and communication links.

Rovers have only been used for the exploration of the surfaces of the Earth, Moon and the planet Mars and so far, only the Unites States and the Soviet Union have successfully implemented rover systems within their missions.

The term *rover* in the context of space exploration refers generally only to the mobile platform, though this is just the last element in a very complex system. Further subsystems needed for a successful rover missions are

- Surface segment providing all necessary subsystems for landing.
- Space segment for transfer and communication purposes.
- Subsystems for the utilisation and the way of implementation on the mission.
- Ground segment for the ground control of the mission.
- Mission operation subsystems (including the control and communication architecture).
- Launch and transfer system elements.
- Subsystems to allow operation in the target environment.

There are several performance drivers for rovers used within space applications. One of the main drivers is the perception capability. This means either the ability to recognise and understand the environment the rover moves through, or the identification of the targets, that they take samples of. In other words the rover has to perceive the near surroundings by seeing it, processing the information they have seen, and analysing what this means for the further action. This procedure has to be done for each segment of movement as well as for obtaining and respectively the analyses of samples.

A further driver closely related to this is the mobility capabilities. The mobility must be sufficient, in order to reach all scientific targets when moving through the types of terrain the rover might encounter. The mobility aspects are also important for the processing, handling and preparation of samples for analyses. A rover will have to enable a close contact between an instrument and the sample. It may have to grasp and collect samples or even have to dig, grind or drill to prepare the sample for scientific analyses instruments.

Especially for mission targets beyond our moon, with long signal delays, a further driver for space rovers is their autonomy capabilities. Rover missions require autonomy to cope with hard real–time situations, but also for limiting the increase of the mission duration caused by signal delays. The need for a high level of ground interaction is also always a cost factor for the operational phase of a mission – however the development of a reliable autonomous system is expensive in the development phase. This demand for autonomy includes autonomous decision making in an unknown and unstructured environment. The rover system has to cope with unexpected situations and has to provide solutions for any contingency. Autonomous operation capabilities implemented in space rovers also simplify the payload (P/L) operations as well as the scientific target selection from the operational point of view.

A rover system equipped with sensors for scientific analyses can utilise these capabilities also for the planning of single operations. Like that the rover can use instruments facilities for mission planning aspects. This use of synergy effects enhances the system efficiency by reducing any additional instrumentation and thus system overhead by a careful integration of the instrumentation into the system.

Autonomy makes the rover adaptable to a variety of situations and missions and provides the necessary flexibility which the systems needs in a completely unknown environment. A space exploration approach in the rover system development can be achieved, if the rover is developed together with the scientific payload in order to optimise synergies.

2.4.1 The classification of rover systems

The classification of space rovers can be typically performed by categorising the system mass [4]:

- A nanorover is typically lighter than 1 kg.
- The mass of a microrover, like the Nanokhod, is in the range of 1 to 10 kg.
- The category minirover is defined by rover with a mass between 10 and 100 kg.

With a mass above 100 kg the vehicles belong to the general rover category.

Typical targets for the different rover systems can be differed by the level of mobility that they can achieve:

- Local Mobility ($> 2 \ldots 50$ m): Overcoming of the lander restriction in order to access locally relevant and characteristic samples.

- Regional Mobility (1...10 km): Access of regionally interesting topographic features like craters, cracks or trenches.
- Global mobility (>100...n×1000 km): Access of global features like mountains, volcanoes and large craters or plateaus. This is currently not feasible on a ground–based exploration.

2.4.2 Design and development of space rovers

The terrestrial use of robotic elements is a growing industry with implementations for a variety of purposes. Miscellaneous institutes and companies focus on the development of robotic systems for the transportation of scientific instrumentation, which is the first step towards the design of space rovers. However, as mentioned before the number of disciplines involved in the rover development for space application is very high. Thus the objective of the institutes and companies can be very different. While some institutions focus more on the development of control concepts, control algorithms and necessary sensor equipment like CNES (Centre National d'Etudes Spatiales), other institutions mainly examine chassis concepts like the DFKI (Deutsches Forschungsinstitut für künstliche Intelligenz) in Bremen or the Swiss institute ETHZ (Eidgenössische Technische Hochschule Zürich), formerly EPFL (Ecole Polytechnique Fédérale de Lausanne).

The development of complete rover systems for space applications is generally divided onto a number of companies and institutes all specialised in different areas. However for small rover systems like the Nanokhod it is possible and necessary to reduce the number of involved partners.

The United States and Russia are the only states which so far have successfully carried out rover missions. Large institutions like the American JPL (Jet Propulsion Laboratory) or the Russian RCL (Science & Technology Rover Co. ltd.) and VNII Transmash (Russian transport Machinery Engineering Institute) built already many different rover systems, additional to the flight rovers in which they were involved. Among those developments there were also several small rover systems, which are in the size range of mini– or microrovers.

Small robotic systems, which already have flown on Mars missions (see section 2.4.3.3), are the two Russian PROP–M rover of the MARS 2 and 3 missions as well as the successful American rover Sojourner of the Pathfinder mission. A further applied system is the hopping rover PROP–F, which was developed for the Russian PHOBOS2 mission to the Mars moon Phobos. Unfortunately, the contact to the Phobos probe failed before it had deployed the PROP–F robot.

More examples of small Russian or American space rover systems (some with an European involvement), which are not (yet) implemented on missions, are the following:

- *IDD 1&2* are two Instrument Deployment Devices developed by VNII Transmash under a contract of MPCh (Max–Planck–Institute for Chemistry). These systems are the precursor systems of the Nanokhod.
- *MRoSA2* is a bigger version of the IDD rover and was developed by RCL together with the Helsinki University under an ESA contract. The increase in size of the system became necessary for the drill system which has been accommodated within MRoSA2.
- *LRMC* is a 4–wheeled Moon rover mock–up, built by VNII Transmash under an ESA contract.
- *MUSES–CN* was developed by JPL for the Japanese Hayabusa mission. However it was replaced by the lighter hopping rover system *Minerva* (see below).
- *Pebbles* is a Mars rover developed by the Massachusetts Institute of Technology (MIT).
- *Rocky 7* is the follow–up model of the Sojourner model developed by the JPL. It is approximately the same size but with higher payload capabilities.

Within the ESA member states the development of space rover systems is much less distinct. In the last years, ESA planned for a few rover missions; however none was conducted so far. As part of these planned missions some rover developments took place in Europe. The involvement in rover developments highly increased when ESA decided to go for the upcoming ExoMars mission. A large part of this development was conducted with the support of the Russian facilities RCL and VNII Transmash.

A few examples of small rover systems developed in Europe are given below:

- *EVE* and *IARES* rovers were developed by CNES using VNII Transmashs Marsokhod platform as a basis.
- *MIDD* (Mobile Instrument Deployment Device) is a small tethered explorer for Moon and Mars operation, developed by DLR and a team which is now at the University of Würzburg.

- *SOLERO* (SOLar–powered Exploration ROver) is an innovative mini–rover for regional mobility on a planetary surface. This chassis concept along with a mission study was developed by a German–Swiss team lead by the company *von Hoerner & Sulger GmbH* under an ESA contract as part of the preparation of the European ExoMars mission [5].

Other rover developments take place e. g. in the Japanese Institute of Space and Astronautical Science (ISAS). The small hopping rover MINERVA (MIcro/Nano Experimental Robot Vehicle for Asteroid) was part of the HAYABUSA (MUSES–C) mission to the asteroid Itokawa, launched in 2003. The small robot lander had a mass of less than 600 g. Unfortunately, after deployment in 2005, the small probe drifted away from Itokawa and did not land on its surface.

Also the upcoming space nations India and China started to develop space rovers. Both nations plan missions (Chandrayaan 2 in 2013 and Chang'e 2 in 2012) to the Lunar Surface which shall carry rover systems.

2.4.3 Examples of rover missions

A rover mission starting from the concept phase to the final flight model is a long, complex and cost intensive venture. This is why only very few rover missions have been launched so far. However, the interest of the scientists to carry out in–situ measurements to learn more about the history of our solar system remains. A number of up–coming or planned missions shall comply with this inquisitiveness. Some of the missions are currently only in initial or concept stages, which means there is still some possibility for re–scoping or cancellation of the mission.

2.4.3.1 Previous Moon rover missions

- Lunochod 1/Luna 17 – the Russian moon rover was launched on the 10th of November 1970. It was the first remote–controlled rover to land on any celestial body. It worked for more than eleven month successfully on the moon surface.
- Lunar Roving Vehicle was an American rover that was used more than once; it was launched with the missions Apollo 15, 16 and 17 in 1971 and 1972. It was used to transport the astronauts on the moon surface. The rover was manually controlled by the astronauts.

- Lunochod 2/Luna 21, the second Russian moon rover was launched on the 8th of January 1973. Like the first Lunokhod the rover was equipped with cameras to take pictures from the lunar surface. Additionally it carried some scientific instruments e. g. to study mechanical properties of the lunar surface material. The mission duration was approximately 5 month.

2.4.3.2 Planned Moon rover missions

- The Chandrayaan–II mission is a joint venture between India and Russia, with a launch date planned for 2013. It consists of a lunar orbiter and a lunar lander with a wheeled vehicle. The rover is equipped with scientific instrumentation for chemical analyses.
- Chang'e 2 mission is the second lunar probe and part of the Chinese Lunar Exploration Programme. The probe is similar to the first Chinese moon probe Chang'e 1 and is scheduled for 2012. This second probe shall test soft landing skills and transport a rover to the lunar surface.

2.4.3.3 Previous Mars rover missions

- A rover was part of the Russian Mars 2 mission – it was launched on the 19th of May 1971. The rover PROP–M has a mass of 4.5 kg moving on ski–walking tractive elements. Unfortunately, the rover along with the lander was destroyed when impacting on the Mars surface.
- Also the Russian Mars 3 mission had a rover aboard. It landed successfully on the 2nd of December 1971, but the lander stopped transmitting only 20 seconds after the landing and the rover never was operated.
- Sojourner was the first successful Mars rover. It was part of the American Mars Pathfinder Mission and landed on the Martian surface on the 4th of July 1997. Although the rover carried a scientific P/L, its main objective was to be a technical demonstrator, e. g. for the Airbag system which was used by the following missions.
- Spirit is the first one of the two Mars Exploration Rover (MER–A). It landed

on the 4th of January 2004 near the Gesuv crater and worked successfully until January 2010. Then NASA redefined Spirit's mission to a stationary research platform, after it got stuck. While the Sojourner rover was in the range of a minirover with 10 kg, the two MER rover each have a mass of approximately 185 kg.

- Opportunity followed as the second Mars Exploration Rover (MER–B) only 21 days after Spirit and landed in the Meridiani Planum. Opportunity is still working successfully (status January 2010), despite the original mission duration being planned as 3 months.

2.4.3.4 Planned Mars rover missions

- The Mars Science Laboratory, named as *Curiosity*, is the next American Marsrover with the launch planned for 2011. The rover system will have a mass of more than 800 kg including a P/L mass of 65 kg. The rover system will be powered by radioisotope thermoelectric generators (RTG).
- The ExoMars rover for the ESA Cornerstone Mission Aurora will be a mobile laboratory having an Exobiology Payload (*Pasteur*) including a geochemical package. As the first European rover mission ExoMars will combine technology demonstration with scientific investigations. A feature of the mission is a drill, which will reach to depths of up to two meters. With the launch planned for 2018 as part of an ESA – NASA cooperation, the development is under progress and the first breadboards are built and tested [6].

2.4.4 Mobility as a key feature for exploration

Current trends in planetary space exploration aim for a mobile system. In order to cover the whole range of scientific interests and to have a well–funded scientific examination of a planetary surface more than one sample has to be considered. This requires a mobile system to reach different areas of interest.

A wide spectrum of results gives a greater view of the explored target. There are two ways to explore a planetary surface: The first option is to design a tool which collects probes and transports them to the lander where they are examined. The other way is to transport the scientific instruments to the sample and carry out in–situ measurements. Either method though requires the ability of movement. The later option though reduces the travel distances and thus increases the area that can be investigated. Additionally the autonomous system can be less complex as the rover will not have to store all digital terrain model information, in order to find back to the lander element.

The mobility aspect of the rover makes it possible to explore a whole region of up to some tens of km around a landing point with a single space system, depending on the implemented system. This regional exploration provides a deeper knowledge and allows a more general characterisation of certain areas of a planetary surface. The mobility of such a system provides the freedom to reach and examine specific points of interest to the scientific explorers. This offers also the opportunity to carry out a systematic exploration of the region. Depending on the system used and the environmental conditions encountered the implementation can in some cases even compensate for the uncertainty of the landing ellipse.

For the mobility in space infrastructure the following development can be observed: For nearby planetary bodies, which have been explored before, the implemented mobile systems are developed with an increasing range (m → km). However for planetary bodies which require a long transfer and which are examined for the first time or for low cost missions, the focus is rather on a system with very low mass. Low mass systems are also considered for missions exploring regions which are difficult to reach or provide harsh environmental conditions, but are of specific interests as e. g. Lunar or Martian polar regions. Mission success in this case is to gain in–situ analyses in the near surroundings of the landing point.

A variety of rover systems have been developed in the previous years in order to accomplish requirements for the exploration of target surfaces. Theses systems differ in their range, terrainability, complexity, their demands on control, power etc.. Thus a rover system can be adjusted to the considered mission scenario.

3 The robotic space system – *Microrover Nanokhod*

The Nanokhod was chosen as model payload for the Mercury Surface Element (MSE) of ESA's cornerstone mission BepiColombo to the planet Mercury. This decision was based on several research and development projects along the microrover Nanokhod, which contributed to the technical maturity of the rover. In preparation for this mission a road map of several projects was defined which foresaw the development of a rover flight model with integrated scientific payload instruments as part of the MSE lander system.

3.1 The Nanokhod rover

The Nanokhod is a small rover system, which fulfils the need of local mobility on a planetary exploration mission to investigate the near surroundings of the lander.

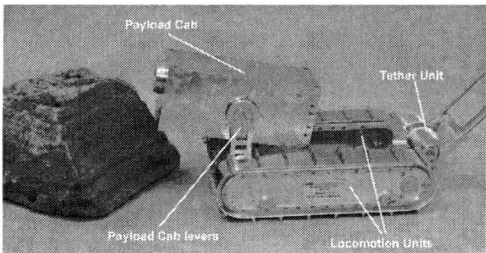

Figure 3.1: Labelling of the Nanokhod rover components

The system is a small and mobile exploration platform carrying out in–situ exploration by transporting and operating scientific instruments to interesting samples beyond the landing point. The scientific payload of the rover consists of two spectrometers and a small camera for the analyses of the chemical and mineralogical composition of the Mercury surface. The payload has to be highly integrated within the rover system such that it meets the given mass and volume requirements (section 3.5) for the rover payload.

Throughout the whole mission on the target planet, the Nanokhod is linked to the lander via a tether consisting of two thin cables. This connection provides the necessary power supply to the rover as well as the transmission medium for the data transfer to Earth, which is carried out by the lander. The tether wires are deployed from pivot–mounted spools that are attached to the tether unit on the rear of the rover. The tether is drawn out with the forward motion of the rover. The locomotion of the rover is solved by two track units. This allows the optimum method of locomotion on the regolith surface for this scale of system.

With a system mass of only 3.2 kg, the Nanokhod belongs to the category of microrover. The system mass includes a payload mass of approximately 1 kg for scientific instruments. The big advantage of the Nanokhod over other rover systems is the use of the lander equipment for power supply and as a data relay. This is part of the approach of a highly integrated payload: The rover avoids unnecessary redundancies and shares generic supporting equipment with the lander. This leads to a payload to system mass ratio which is unequalled by any other autonomous rover system (see Table 3.1). Of course the linkage to the lander limits the exploration range to the near surroundings of the landing point. However, this is completely adequate for a mission with the main mission goal to collect the first in–situ Mercury surface data like the BepiColombo MSE. The baseline mission requires an operational life of two weeks; the extended mission is two additional weeks of operational life. The relatively short mission duration is imposed by the difficult conditions on the Mercury night side with limited energy resources which do not allow for an exploration of a larger region. A mission, which is significantly restricted by available resources, is ideal for a rover system like the Nanokhod, which is based on pan–system synergies to create an optimal highly integrated system.

Rover system	Rover System mass [kg]	Included mass of scientific payload [kg]	Payload to system mass ratio
MER	185	6	0.032
MSL–09	775	70	0.090
Sojourner	11	0.7	0.064
ExoMars	190	30	0.158
Nanokhod	3.2	1	0.313

Table 3.1: Payload to system mass ratio of various rover systems

The Nanokhod system with its high ratio of payload to system mass and the low overall mass and the small volume is thus perfectly suited for either low cost missions or missions to demanding far distant targets, which both have to cope with a strongly restricted mass and energy budget.

A system trade–off between the microrover and a robotic arm on the lander results into a similar system mass. But the rover has an additional degree of freedom due to its mobility. This is enlarging the operating range for scientific exploration by far.

The Nanokhod has quite impressive climbing abilities, being able to overcome obstacles higher than its height and cross trenches of up to 10 cm. This can be useful for the deployment of the rover from the lander but also for the recovery from unforeseen situations (e. g. tipping over of the rover). These climbing capabilities might even be useful to overcome rocks, although this would be complex and require a high control effort. However for the Mercury surface rocks are considered to be rare and thus are rather seen as exploration targets than as obstacles. Rocks are the scientifically most interesting objectives expected on the Mercury surface. Nevertheless, the moving and climbing abilities of the rover allow for a flexible adaptation to any situation which the system might encounter on the landing site.

The Nanokhod rover system consists of three segments (Figure 3.2): The mobile segment, which is the rover itself, the lander segment for communication purpose with the rover, as well as the Ground Control Segment, the rover control centre on Earth.

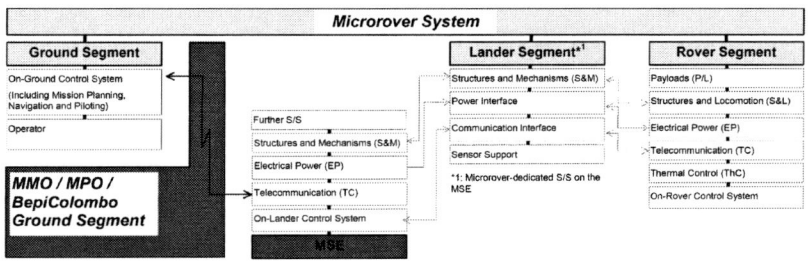

Figure 3.2: Segments of the rover system

This thesis deals mainly with the subsystem rover and the system parts integrated on the lander, which shall be defined as the new main system. This is sensible, as the MRP project (Mercury Robotic Payload – see section 3.3) focuses on the development of the rover, while the other components, like the lander and the BepiColombo ground segment, are not further developed with the MSE being cancelled in November 2003. The influence of their boundary conditions and of the interfaces to the rover system was nevertheless taken into account by defining a detailed set of requirements (see section 3.5) in the forefront of the project, which the rover design had to meet and fulfil.

3.2 Development history of the Nanokhod rover

The Microrover Nanokhod[4] bases on a concept of the Russian institute VNII Transmash. The development of the rover was then continued in a cooperation with the Max–Planck–Institute for Chemistry, Mainz with the support of the DARA for the MARS98 programme. Finally in 1992 the Max–Planck–Institute transferred the development of the rover to the company *von Hoerner & Sulger GmbH*.

Figure 3.3: Historical Nanokhod rover models

vH&S continued the development of the rover in the following years considering several different design aspects and for a variety of mission scenarios. The rover made significant progress to soon provide a practical flight implementation. The design steps included e. g. the accommodation of scientific payload instruments [7]. A drilling tool was developed for a larger version of the Nanokhod [8]. Prototypes were built during the development to test mobility performance like step and slope climbing [9]. Sealing concepts were implemented to protect

---

[4](NANO–KHOD: Greek *nano* [ναννωσ] =dwarf, Russian *khod* [ход]=walker)

the rover's inside from regolith and have been tested in detail. During its development the rover has been targeted for different planetary bodies such as Mars, Moon and Mercury. The last e. g. considered both day and night side conditions and thus extremely hot and cold temperatures [10].

The latest development of the rover is now a Nanokhod design which is able to cope with the challenging requirements of a flight model on the Mercury night side. A hardware model of this design is realised for environmental tests such as vibration, shock and thermal vacuum to the extreme requirements of the Mercury mission profile (section 3.4).

3.3 The ESA research project: Mercury Robotic Payload – MRP

The MRP project aims to explore the impacts on the rover design, which are related to extreme conditions of the Mercury environment. The mechanical aspects of the MRP rover model are developed to the level of an engineering model. The electronic design is also developed for the operation under the Mercury night side conditions, although the project resources did not allow to build a full flight implementation with respect to the utilisation of radiation hard/flight level components. The main goal of the project was the realisation of a feasible design for the required mission conditions; e. g. with the thermally demanding environment one of the main efforts was the demonstration of a thermally compatible system. The experience gained at this stage shall significantly reduce the risks of the final flight design, [11] [12] and [13].

Despite the cancellation of the BepiColombo lander for which the rover was developed, the basic conditions and mission requirements were kept the same. Although currently there is no specific mission for the rover, the work on this rover project for the Mercury conditions and the experiences gained are highly relevant for upcoming missions. The testing of such a complex and highly integrated system dedicated for a demanding environment has a valuable output for future applications. The rover gained a maturity within this project, which allows a fast and cost effective implementation for an alternative mission. The project moved the rover system from a study level to an engineering level for demanding mission requirements and included the integration of three payload instruments. It can be thus summarised that the project prepares the rover design for a variety of missions and their requirements.

3.4 The ESA cornerstone mission BepiColombo

The BepiColombo mission to Mercury proposed in 1993 was chosen by ESA's Space Science Advisory Committee (SSAC) to be the 5th Cornerstone mission of ESA. The mission was named after Giuseppe ("Bepi") Colombo from the University of Padua, who first explained the unexpected resonance of the Hermean gyration. This mission, which was planned for a launch in 2009 (now postponed to 2013), originally consisted of three elements [14]:

- The Mercury Planetary Orbiter (MPO), equipped with an instrumentation for remote sensing, radio science and the observation of asteroids.
- The Mercury Magnetospheric Orbiter (MMO), equipped with sensors and instruments for the examination on electric fields and particle distribution around Mercury.
- The Mercury Surface Element (MSE), a lander unit equipped with instrumentation, allowing for in–situ analyses of mineralogical and chemical properties of the Mercury surface. However in the course of the MRP project the MSE has been cancelled from the mission for cost/feasibility reasons.

The BepiColombo mission foresees two launchers; one for the MPO and one for the MMO (former MMO and MSE).

The transport of the elements to Mercury is realised by a combined thrust system using electrical and chemical propulsion as well as gravity assist manoeuvres. The interplanetary transfer uses a Solar Electrical Propulsion Module (SEPM), which is jettisoned upon arrival. For the insertion in the polar Mercury orbit the orbiter elements use a Chemical Propulsion Module (CPM), which is also jettisoned as soon as all elements are fully deployed. For the descent of the lander it was planned to use chemical propulsion and airbags.

The mission is a cooperation between the European Space Agency (ESA) and the Japanese Space Agency (ISAS). The scientific targets defined for the BepiColombo mission are the following [15]:

- Examination of the origin and the evolution of a planet so close to the central star.
- Examination on the morphology, the inner structure, the gyration, the geology and the composition of the craters of the planet.
- Examination on the origin of the Mercury magnetic field.

- Examination of the dynamic behaviour and the composition of the Mercury exosphere.
- Examination of the structure and the composition of the Mercury magnetosphere.
- Examination of the interplanetary medium in the Mercury orbit.
- Verification of Einstein's theory of relativity.
- Detection of potential Earth endangering asteroids.

3.4.1 The planet Mercury

The planet Mercury was probably already seen by prehistoric men, but the first recorded observation is by the Greek astronomer and philosopher Timocharis of Alexandria in 265 B.C.. Since then the planet has been observed by all available means and played an important role in the religious life of many ancient civilisations [16], [17]:

- Greeks first regarded the planet as two objects of which one was seen in the evening – Hermes (evening star) – and the other one in the morning – Apollo (morning star). Later when they recognised the planet to be only one object they called it Hermes, being the god of twighlight and dawn and the messenger of gods. The current name *Mercury* derives from the Latin god Mercurius which is the Roman designation of Hermes.
- Egypts called the planet Sabkou and were the first to observe that the planet orbited the sun.
- Teutonic people referred to the planet as Woden, the chief god in the Norse mythology.
- In 1630 Kepler first predicted the Mercury transit by the sun. The transit was then observed by the French philosopher, scientist, and astronomer Pierre Gassendi on the 7th November 1631.
- In 1639 the Italian astronomer Zupus and independently in 1644 the Protestant councillor and mayor of Danzig, Hevelius, made the first observation of the Mercury phases.
- Johann Hieronymus Schröter (August 30, 1745 – August 29, 1816) and Karl Ludwig Harding (September 29, 1765 – August 31, 1834), both German

astronomers, discovered the first surface markings on the Mercury surface.

Figure 3.4: Mercury, picture from the NASA Mariner 10 mission

Although Mercury is one of the planets which has been known for so long, it is still today relatively unexplored. Due to the planets vicinity to the sun it is difficult to observe from Earth. It is never more than 28 angular degrees from the Sun as viewed from the Earth and can only be seen in the twilight of the early morning and the late evening. In both cases there is a long optical path through the Earth atmosphere, which is disturbing any observation.

Observations from the Earth orbit, where disturbances of the Earth atmosphere are minimised, are also not much more preferable. The instruments for the examination of Mercury have to be pointed more or less in the direction of the sun and such observations bear a great risk of damaging the instruments by intense solar radiation.

The use of a Mercury orbiter for observation is difficult as the transfer to the planet is also a demanding task. This is mainly due to the high gravitational potential of the sun when travelling from the Earth to Mercury, which is requiring a high amount of propellant for deceleration. In order to reduce the required amount of energy, the trip can be realised with an advantageous planetary constellation using the gravity assist effects to decelerate, which though has the disadvantage that it requires a longer time. The extended time aspect has then to be considered in respect to the life time of components e. g. considering the radiation hardness of the system. While the space system approaches Mercury and thus the sun, it is exposed to increasing solar and heat radiation. Note that the radiation from the sun is inversely proportional to the squared distance between the sun and the space probe. When finally entering the Mercury orbit, the satellite is exposed to high thermal variations resulting from the night and day side of the planet.

Mercury has a radius of 2440 km which is only slightly bigger than the Earth moon. The Mercury orbit around the sun is very eccentric, with a distance to the sun varying between 0.308 und 0.446 AU. It needs ~ 88 days for one revolution around the sun and about 2/3 of the time (~ 58.6 days) to rotate around its own axis.

Being one of the terrestrial planets, Mercury (together with Venus, the Earth and Mars) carries essential information on how this group formed. The evolution and origin of the planets is the key to understand how the conditions for life could develop. Especially since the high density of Mercury varies from linear behaviour of the ratio size to density, which is given by the other terrestrial bodies.

The environmental conditions of Mercury [18], [19] represent a great challenge for any space system. The expected surface temperature vary in between $-180°C$ on the night side of the planet and $427°C$ on the day side of the planet. The tenuous atmosphere of Mercury is almost negligible. So far five atmospheric elements have been identified: Atomic oxygen, atomic hydrogen, helium, sodium and potassium. The atmosphere is a very rarefied medium so even the neutral constituents of the surface gas almost never collide, which technically makes the Hermean atmosphere an exosphere. However, the gas–surface–interaction is completely different from the interface between the Earth exosphere and atmosphere. The total pressure is expected to be ~ 10^{-15} bar. The surface of Mercury is covered by regolith with grain sizes of $10 - 100\ \mu m$, similar to the moon surface in the lunar mare [20]. This is caused by long term gardening effects, such as bombardment of micrometeoroids or erosion by solar wind.

3.4.2 Previous Mercury missions

So far, the only accomplished mission to Mercury was the NASA probe Mariner 10 which was launched on November 3, 1973. The transfer to Mercury including a Venus gravity assist manoeuvre took nearly half a year. This mission trajectory, which allowed for the less powerful and thus cheaper Atlas–Centaur carrier, was planned by the Italian mathematician Giuseppe Colombo.

The main mission objective for the robotic space probe was to investigate the environmental conditions of Mercury and Venus. A secondary objective was the investigation of the interplanetary medium. In its lifetime, Mariner passed three times the planet Mercury with a distance varying in between 327 and 50000 km, however still many important questions were left open.

3.4.3 Current and planned Mercury missions

The scientific interest on Mercury is still very high, as so far little is known. In addition to ESA's BepiColombo mission, which is described in section 3.4, the Messenger mission is the only other current and future mission towards Mercury.

3.4.3.1 Messenger

The current American mission MESSENGER (MErcury Surface, Space ENvironment, GEochemistry, and Ranging) executed its first Mercury flyby in January 2008, the second in October 2008 and the last and third flyby in September 2009. These flybys allowed already to gather first information before the probe will enter the Mercury orbit. The data analysed up to now indicate that volcanoes were involved in the formation of Mercury plains. The first results from the data about the planets core suggest that the magnetic field is actively produced in the core. More results will follow with further evaluation of the gathered data.

The probe was launched in August 2004 and will be the first artificial object to enter the Hermean orbit in 2011. One of MESSENGERs objectives is to fully map Mercury for the first time. From the orbit the probe shall investigate the geological and tectonic history of the planet as well as its composition. Also the magnetic field of Mercury and its origin is of major interest and will be explored. Further aims are the investigation of the planet core size, the polar caps and the Hermean exosphere and magnetosphere.

3.5 Requirements for MRP

The MRP project requirements are defined to meet the BepiColombo mission requirements [15] and the BepiColombo Payload Definition Document [21] and they are derived from the requirements already defined for the previous project RTPE (Rover Technology for Planetary Exploration) [22].

The MRP project was performed in parallel with the closely related TRP study GIPF (Geochemical Instrument Package Facility) [23], which is responsible for the integration of the scientific payload into the rover; the requirements for the two projects were defined together in the same database to view their interrelations and dependencies.

The requirements defined for MRP and GIPF are imposed by an external entity towards the system considered in the project. These external entities can be subdivided in the following two main categories:

- User requirements:
  The user for these projects is the principal investigator (PI) or the team of scientists, who is/are planning to do scientific investigations with the MRP and GIPF instruments. Thus the requirements defined under the segment "user" focus on the compliance of all systems to produce scientific output. With more than one PI or instrument, there is also more than one scientific goal, each with its own set of requirements.
- Mission requirements:
  The mission requirements are requirements imposed onto the system by the overall mission to achieve the mission goal. These requirements affect the whole technical, physical and organisational environment to which the systems are exposed. These requirements thus add up from a variety of different entities, which are:
  - The landing site with the harsh Mercury environment has direct and indirect influence on the system and all component subsystems, e. g. from thermal and radiation aspects.
  - The spacecraft with its requirements for the accommodation and conditions during the transfer, e. g. contamination aspects towards other instruments accommodated on the lander.
  - The lander configuration to which the rover is connected and with which it is interacting for data exchange and power supply.
  - The technical requirements needed in order to provide the mobility required from the user to do their scientific investigation.
  - The ground segment where the operational planning is done and which will provide the Earth link for the rover system.

Furthermore the requirements are classified into two degrees of necessity, which are:

- Mandatory (M): Requirements which are absolutely necessary to be implemented in order to assure mission success. These requirements are always formulated using a "shall".
- Desirable (D): The implementation of the desirable requirements is not absolutely necessary, though if implemented it will be a valuable contribu-

tion to the mission. These requirements often refer to the extended mission and are formulated with a "should".

If a desirable requirement conflicts with a mandatory one, the mandatory requirement gets the priority in implementation. Desirable requirements may be skipped due to various reasons, e. g. limitations in complexity, cost or implementation risk.

3.5.1 Main requirements/drivers for the MRP project

The full list of the project requirements can be found in [24] and [25], however the requirements with the highest impact on the design shall be summarised in the following paragraphs.

The main drivers for the Nanokhod design during the MRP project are the constraints in mass and volume. Most other requirements are related to these issues, which adds additional difficulties in their accomplishment.

The major requirements towards the rover system imposed by the mission are the following:

Due to the energy intensive transfer to Mercury, the payload mass shall be minimised in order to reduce the need of propellant and thus to make such a mission feasible. This is why the rover system mass, including the scientific payload, is strictly limited. With the mass of a structure being dependent of its volume, the rover system is also very restricted in its size. This implies for the Nanokhod design that the rover system shall be designed to have a maximum mass of 3200 g and the rover segment shall not exceed the stowed volume on the MSE of $240 \times 165 \times 65$ mm^3.

The mass limitation also drives the design of the power system. As the lander is targeted for a night side landing on Mercury, the power for the whole mission has to be supplied chemically. Thus in order to reduce the battery mass, the power consumption of the rover system shall be minimised, both in overall power consumption and peak power. The rover system shall not draw more than 6 W peak power from the lander during operation and shall consume maximum 265 Wh of electrical energy during the baseline science mission including deployment operations. This maximum power/energy also includes the power for controllable heaters, which may be needed for the preheating of the drive electronics.

With a transfer phase duration to Mercury that can last up to 4 years, the solar radiation will have quite an impact on the design. Due to mass limitations, the probe will limit additional structure for the radiation protection of all payloads.

This means that the requirement for radiation protection is passed on to the rover so every payload shields itself to the required level. For the Mercury mission, the rover shall be designed to withstand a charged particle radiation dose of 22 kRad(Si).

The MRP–Nanokhod is designed to land on the Mercury surface as part of the BepiColombo MSE. Due to the lack of atmosphere, the deceleration of the lander has to be performed by using chemical propulsion. In order to reduce the amount of chemical propulsion to a minimum, the rover shall survive a landing shock up to 200 g for a duration of 20 ms without degradation.

Initial proposals for the rover mission considered both day as well as the night side conditions of Mercury, the landing site was eventually decided to be only on the night side of the planet. This has the advantage that the thermal range is reduced due to the absence of the sun's radiations. The reduction of the thermal range is the only way to realise a mission to Mercury, which is highly restricted in mass and power.

The night side landing becomes possible with the long sidereal Hermean day which lasts for 88 Earth days. Despite the reduced thermal range, the environmental conditions remain severe. The Mercury night side couples a high vacuum with surface temperatures estimated to be down to $-180°C$, which shall be survived by the rover system. In addition to the surface requirements the rover system shall withstand non–operational temperatures of up to $+70°C$. The constraint in the energy and mass budget mentioned above also implies that the harsh temperatures can not be balanced by an active thermal control system, which would be far to power– and mass intensive.

The rover system shall survive a baseline mission duration of 14 Earth days with the possibility for a mission extension of another 14 days (desirable mission extension). The lifetime of the rover mechanism is mainly endangered by the fine regolith surface. The system components, like e. g. the drive system, bearings but also the electronics, shall thus be protected to ensure mission success.

The main requirements imposed to the rover system by the scientist are as follows:

The regolith not only imposes mission requirements but also requirements from the user side on the system; the rover shall be able to move across the Mercury surface, which is expected to consist of fine regolith similar to the Moon surface. The system shall provide the abilities to overcome slopes of up $20°$.

Although it is expected to encounter mainly a heavy gardened surface, it is also required that the rover shall be able to negotiate steps of 10 cm height and trenches of 10 cm width. This requirement is not only for the negotiation of obstacles, but also for the recovery of the rover in case it tips over. In both cases the rover is using the payload cabin drives to lift the locomotion units. For the obstacle negotiation the PLC is moved on the obstacle and the LUs are lifted up onto the obstacle. For the rover recovery the PLC stays on the ground while the LUs are lifted up and moved by 180°. A further mandatory requirement to the drive system is the maximum speed which is set to 5 m/h.

The rover communicates and is powered by the lander via tether cables which shall be at least 50 m in length. This length shall also allow the rover to explore an area of at least 5 m radius from the landing point. The limitation in exploration radius is mainly driven by the constraint that the rover control system utilises the stereo camera of the lander and the rover is thus limited to the field of view of the camera.

For situations with obstacles that the rover cannot overcome, it shall have the ability to drive backwards for one full rover length without crossing the tether wires and to carry out spot turning of ±90°.

The main mandatory scientific requirement is the accommodation of three scientific sensor heads located inside the rover payload cab. Related to the accommodation of the instruments, a whole set of requirements are necessary for the positioning and then the operation of the instruments. For example the mission shall carry out one in–situ analyses per Earth day with each of its three instruments: Microrover camera (MIROCAM), Alpha–Particle X–ray Spectrometer (APXS) and Mössbauer Spectrometer (MIMOS). This means the rover shall provide the possibility to position all sensor heads on the same sample, no matter if it is on the flat surface or on an inclined obstacle surface. The Mimos instrument additionally requires contact force of 1 N, in order to produce reliable measurements, as the MIMOS core is oscillating. This shall be supplied by the rover segment.

Due to a single communication period once an Earth day, there is a need for autonomy of the rover system. The autonomy shall be supported with the stereo camera of the lander, which creates the environmental model in which the rover operates. The near–autonomously operation of the rover implies the need of the system for several sensors, which provide the system enough information not only about the own state, but also of the surrounding area.

4 Design of the Nanokhod rover for the Mercury environment

The general configuration of the Nanokhod rover is shown in Figure 4.1. It is a tracked vehicle consisting of three main bodies. The centre body is the payload cabin (PLC), which contains the scientific payload instruments. On each side of the PLC a tracked unit contains electronics and provides the locomotion of the rover – the locomotion units (LU). The LUs are rigidly connected with each other via the tether unit (TU) in the rear of the rover vehicle. The payload cabin is mounted on the front ends of the locomotion units with two lever arms. This configuration provides two degrees of freedom for the payload cabin that allow the positioning of the scientific sensor heads. The arm on the right side of the PLC is used as a feed–through for the supply cables into the PLC. The left arm is mounted with each end on the two PLC articulation drives in order to allow lifting and rotation of the instrument housing. This articulation of the PLC can also be used to overcome obstacles such as trenches or steps and to recover in the case the rover tips over.

The volume of MRP rover in stowed configuration is $232 \times 162.2 \times 67.4 \, \text{mm}^3$, with an overall mass of 2904 g including the mass of the instrument heads of 880 g. The rover has a peak power of 5.7 W at the lander side for the simultaneous activation of two drives; either for the rover locomotion or for the positioning of the payload cabin. This value includes the heater power required to preheat the drive electronics.

4.1 General aspects of the Nanokhod design for Mercury conditions

The project requirements (see section 3.5 and [24]) which fulfil the BepiColombo mission requirements define a challenging framework for the design of the rover system.

The general rover configuration is based, as far as possible, on former Nanokhod developments. With the strict volume requirement the outside dimensions of one

Figure 4.1: MRP configuration of the Nanokhod rover with a Euro coin for a size
comparison

body could only be increased if the dimension of another body is decreased cor-
respondingly. The configuration of all bodies is driven by the rover's operational
principle: The arm length has to provide the rotation of the deployed PLC and at
the same time the accommodation of the PLC between the LUs within the given
volume. The rover configuration also has to provide a weight distribution which
allows the movement of the PLC far to the front to allow for rover movements
that require the use of the PLC as an extra limb, e. g. for climbing activities.

The structure of the rover system has to be very robust and sturdy to resist
the vibration loads of the launch and the shock loads of the Mercury landing
procedure. At the same time the overall mass and volume has to be reduced to a
minimum. This requires careful design and analyses of the rover structure.

A further aspect which has to be considered simultaneously is the thermal
design. It has to be assured that all rover components withstand the temperature
range of maximum $+70°C$ during the transfer down to $-180°C$ on the Mercury
surface without degradation. This includes thermal cycling effects that occur
on the transfer to Mercury, when components are exposed to varying solar
radiation depending on the orientation of the spacecraft. Further thermal effects
result from the fact that the rover system will be manufactured, assembled and
initially tested under room temperature conditions and will have to function in
an environment with temperatures of $-180°C$. This working temperature range
imposes a careful choice of materials and components to be used for the design.
All electronic components have to be qualified for this range, as there are no
qualified off–the–shelf components available for these temperatures. The design
also has to consider the thermal expansion coefficients of the applied materials,

as tolerances and fittings will be affected.

Also the vacuum environment requires consideration during the design phase. Care has to be taken with the choice of material, e. g. movable metal interfaces have to be designed such that no cold welding effects can occur. The vacuum conditions also affect the thermal design of the rover, as the lack of atmosphere excludes thermal exchange via convection and conduction with a surrounding gas. This is especially critical for the design of the electronic PCBs, the choice of the rover actuator, as well as the thermal linkage of these components. A further aspect of the vacuum is the outgassing effect of materials, which cause problems, especially for optical instruments. Sublimation effects require the use of dry lubricants within the drive units.

From the perspective of the mission requirements, in addition to the design needs evoked by environmental conditions, the main criterion for the design is the accommodation of the instrument heads. The rover payload cabin has to provide sufficient volume for the instruments and take care of all imposed conditions by the rover payload. The focus lies here on the transport and the positioning of the instruments and its accuracy.

For the autonomous operation of the rover, it must have a set of sensors, needed to provide the necessary information for its control. This includes the combination of the camera and a structured light source.

The power supply also has significant implications for the rover design. The Mercury night side temperatures for example cannot be balanced by an active thermal control system. In addition, to further reduce the overall power demand of the rover, the electrical system has to be optimised so that the power needs of the rover are reduced to a minimum, particularly considering the duration of the different activities.

To make the system radiation resistant, radiation hardened components up to the required radiation level or shielding by the surrounding structure have to be implemented. For the design of the current hardware model no radiation resistant components were applied due to financial limitations within the projects. However, the electrical design was conducted bearing in mind the radiation issue and a route to radiation compliant design was identified.

For the accommodation of the rover system within the lander, interface pos-sibilities with a hold–down–device (HDD) had to be considered. The HDD is part of the lander and secures the rover during the transfer to the planet's surface with the means of deployment.

4.2 Locomotion unit design

The two locomotion units (LU) provide the rover's ability for horizontal movement. At the same time the LUs accommodate a significant proportion of the rover electronics. They consist of two parallel aluminium walls that are rigidly connected with two yokes per unit. The yokes are also used to mount the two locomotion motors and, in the left LU, one of the PLC articulation motors.

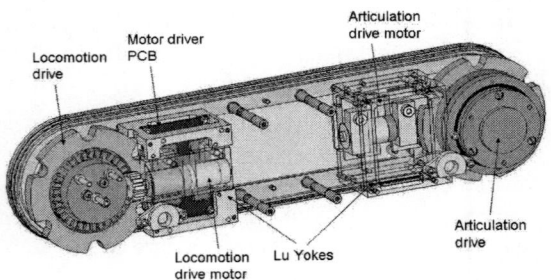

Figure 4.2: CAD model of the left locomotion unit

The output of the locomotion drives in each LU is a sprocket wheel which engages with PEEK (Polyetheretherketone) guides that are attached to the inside of the thin steel track foil. The PEEK guides have steel pins on each end which run in grooves on the locomotion unit side walls.

The steel foil encloses the locomotion units and, when driven, supplies the locomotion movement of the rover system. Aluminium grousers are mounted on the outside of the track foils on the opposite side of each PEEK guide and provide both the necessary grip during climbing activities and additional drawbar pull when driving on a regolith surface. In the front end of each LU the tracks are guided by a free wheeling sprocket. The use of tracked locomotion units provides superior performance over wheeled concepts on dusty terrain for a rover of this scale.

The design of the two locomotion units was kept to a large extent the same for both sides, in order to minimise the parts count and reduce the complexity and cost. This is particularly true for the components of the locomotion drives which are inherently the same. These are located in the rear end of each unit, mirrored to each other. The locomotion motor and its drive electronics are mounted in the rear yoke of the LUs. In between the yokes in the centre of the LU the motor controller electronics are mounted to the side wall on aluminium stand–offs. The stand–offs do not only provide the support for the boards but also a good thermal linkage between the heat dissipating PCBs and the casing of the rover.

However each LU performs different additional functions:

The left locomotion unit accommodates the PLC articulation drive in the front end, which moves the lever arm. Thus it also contains the additional equipment of this drive unit. Integrated in the front LU yoke is the motor and the corresponding motor driver. The drive components for this drive are similar to the locomotion drives, except for the output which drives the arm.

The right LU contains one drive less, but accommodates the electronics of the Tether Data Interface node adjacent to the motor controller board. The right locomotion unit also provides the conduit for the cabling into the payload cabin through the lever arm. A partition in the front of the right locomotion unit provides space for the nine wires of the internal rover bus to be wound into a spiral providing the necessary freedom in the cable length to rotate the payload cabin arms through an angle of 370°. The rotation of 370° provides the deployment and all necessary articulation of the PLC.

4.2.1 Track guide blocking

During the initial testing of the manufactured rover occasional blocks occurred in the locomotion units between the PEEK track guides and the sprocket.

The initial track foil design was made with a single pin on each side of the track guides, alike the design of former rover models. This manner of guiding the track foil allowed the grouser to rotate around the pins and, with high loads on the grouser, causing the deformation of the track. The blocks only happened after one of the grousers was excessively tilted deformed the track foil plastically. This allowed the grouser to remain at an angle and to cause the blocking. It should be noted that the tilting of the grousers must be avoided also to prevent the sealing from failing, which happens when the track foil lifts from the PTFE (polytetrafluoroethylene) membrane.

The deformation in the new design, which was not observed in former designs, became an issue due to the implementation of a new steel alloy for the track foil. The steel alloy X10CrNi18–8 (Type 1.4310) was chosen because it provides elasticity for the low temperature range down to $-180°$C. This elasticity of the track foil was necessary to still provide secure sealing of the locomotion units. The new track foil for the MRP rover also has only half the thickness of the track foils from previous rover models, in order to reduce thermal deformation at low temperatures. This of course means the track foil has a smaller bending modulus and the track foil bends more easily.

During room temperature conditions even a slight bending of the track foil causes plastic deformation. Thus the track foil deforms when carrying out e. g. step climbing operations. For this manoeuvre it can happen that the rover load applies on only one grouser per locomotion unit causing a torque load on the track foil. With the deformed track foil the track guides do not engage properly with the sprockets anymore and thus start blocking in between the sprocket and the track guide grooves (see figures 4.3).

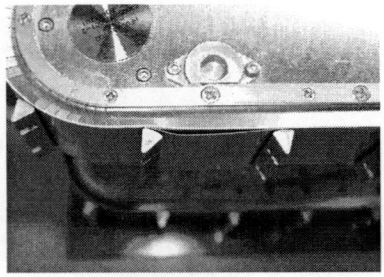
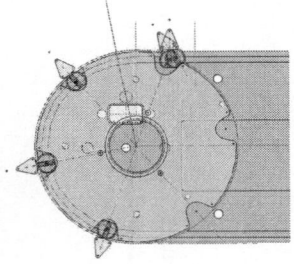

(a) Jammed locomotion unit   (b) Rotation of track pin

Figure 4.3: Initial problem of the track guide jamming of the locomotion units

Another factor increased the probability of blocking with the bent grousers on the track foil. A chamfer was added on the PEEK guides of the MRP design in order to have sufficient volume for the accommodation of all necessary equipment inside the locomotion units. This flat surface on the track guides leads more easily to a mismatch with the sprocket wheel. However this can only occur when the track foil is already deformed and the grouser is tilted to an angle which is causing the seal to fail.

In order to solve the blocking problem and to avoid the seal failure, several different solutions were analysed:

A redesign with smaller chamferless track guides and a matching sprocket allowing for the same space in the LUs was discarded due to cost and time issues. Furthermore this would not fix the problem, as the smaller track guides can still rotate so far that they jam also without a chamfer and also the sealing cannot be ensured.

Another solution was to change the track guide material to one that does not deform so easily for room temperatures. During the design phase, it was also discussed to use CuBe (Copper Beryllium) for the track foil, as it would provide sufficient flexibility for both room and the deep temperatures. However, the bending behaviour for the temperature range would have to be verified to be sure that the blocking problem can definitely be avoided. This solution would thus have been time consuming and cost intensive. CuBe material has additionally the disadvantage that it is very difficult to weld the track foil ends together.

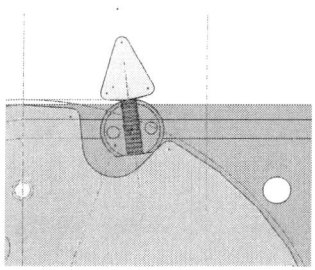 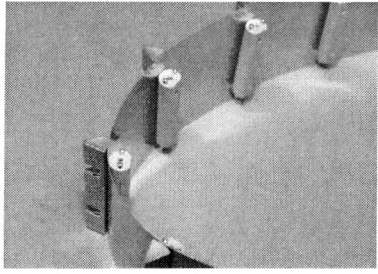

(a) CAD model of track guide with two–pin solution

(b) Track foil with modified track guides

Figure 4.4: Integrated two–pin solution for track jamming problem

The solution that was finally implemented was a modified track guide with two guiding pins on each end of the PEEK cylinder as shown in figure 4.4. The two pins along the line of the groove in the LU side walls allow only a minimal rotation of the track guides, thus allowing only tilting of the track foil in the elastic range. Before implementing this solution, it was verified that the friction value will not cause excessive load on the drives during locomotion or climbing activities. This solution of the blocking problem has the significant advantage of ensuring correct sealing of the locomotion unit.

4.3 Sealing concept of the locomotion unit

The fine dust of the Mercury regolith requires a good sealing concept for the rover system to protect the accommodated electronics, instruments and the drive systems. This is especially true for the locomotion units with the moving track foils, which are difficult to seal.

The sealing of the locomotion units is realised with a circumferential 0.25 mm PTFE membrane between the track foil and the LU side walls. The PTFE foil is mounted on the outside of the LU side wall and is pressed by a pre–stressed spring to the moving track foil.

During the development history of the Nanokhod several sealing concepts have been realised and tested. The sealing was always customised to the mission conditions: For the Mars design the locomotion units were sealed with the help of brush material in between the track foil and the side wall. This could not be used for the Mercury design as brushes would become stiff at the cold temperatures and thus would not properly seal anymore.

PTFE generally stays flexible at low temperatures, and therefore was used for the sealing concept for the previous development stage (RTPE – Robotic Technology for Planetary Exploration). At this stage a commercial PTFE gasket was utilised with an integral steel spring. Although the solution remained flexible at deep temperatures it did not perform satisfactorily at $-180°$C. This was because the concept did not compensate for the different coefficients of thermal expansion between PTFE, the spring and the steel track foil, resulting in the lifting off the PTFE seal from the track foil. PTFE shrinks 10 times more than steel does and with the MRP temperature range varying between $+70°$C and $-180°$C the shrinkage effect becomes significant and directly affected the sealing quality.

For MRP, the spring and the PTFE membrane design was reviewed and the sealing concept was changed, so the shrinkage of the PTFE foil does not cause a gap in between the track foil and the sealing. Instead the PTFE is implemented in the design so that the shrinkage causes only a slight decrease of the contact surface between track foil and PTFE.

The shape of the locomotion unit requires a customised spring in order to provide sealing along the whole track foil. Tests and analyses were performed to define a sufficient spring force over the temperature range. This resulted in the following configuration: The spring is fabricated by etching the shape of the sealing circumference from a sheet of spring steel (Type 1.4310). This spring features fingers in the apexes and a single stripe along the bottom and the top edges of the locomotion unit, figure 4.5. The shaped steel foil is then bent to the predefined angle of $22.5°$ by a customised tool. In this manner the spring provides a force of 0.2 N/m on the PTFE. This value was found to be optimal for the MRP temperature range, with a satisfactory sealing performance and tolerable friction values.

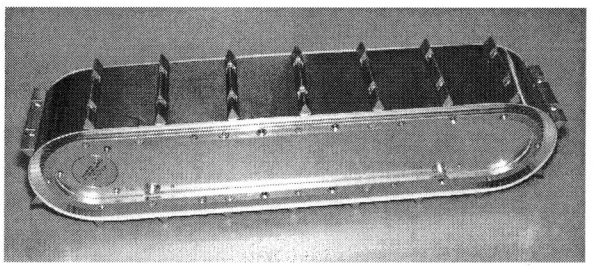 

(a) Sealing on LU side wall   (b) Detail of LU apex

Figure 4.5: Locomotion units are sealed against regolith and particles with a thin PTFE foil which is pressed to the track foil by a customised steel spring

4.4 Design aspects of the central payload cabin of the Nanokhod

The main role of the Nanokhod payload cabin (PLC) is the accommodation of the scientific payload instruments, which analyse the target surface. Due to that the instrument heads are the design drivers for the PLC shape and size. In the framework of the MRP project it was designed to supply all necessary interfaces for the three instrument frontends of the Geochemistry Instrument Package Facility (GIPF – [23]). This instrument package was chosen as the baseline scientific payload for the Mercury mission. GIPF was another ESA TRP study carried out in parallel to the MRP rover development at the company *von Hoerner & Sulger GmbH* with the support of the responsible PIs. The package contains two spectrometers and one small camera which are described in detail in section 4.10. For the accommodation of the payload, the overall system also has to account for the instrument mass and its influence on the system centre of mass (CoM) and the drive design.

The payload cabin is divided into two partitions, the instrument partition described above and the rover control partition as shown in figure 4.6. In such a small system it cannot be completely avoided that the two parts interfere with each other. However, in order to provide flexibility for alternative payload instruments it is desirable to keep them separate as much as possible.

The rover control partition of the payload cabin accommodates the central

(a) Instrument partition with Instrument mass dummies

(b) Rover conrol partition

Figure 4.6: The rover payload cabin is divided into a rover control partition and an instrument partition

on–board controller electronics and the payload cabin control electronics. To allow the rotation of the payload cabin and thus the positioning of the three sensor heads, the rover control partition also contains the second PLC articulation drive unit. Furthermore a stripe laser is accommodated in this partition of the PLC, which is required by the control concept. Together with the camera it is used to generate a simplified model of the surrounding area (see section 4.9.1).

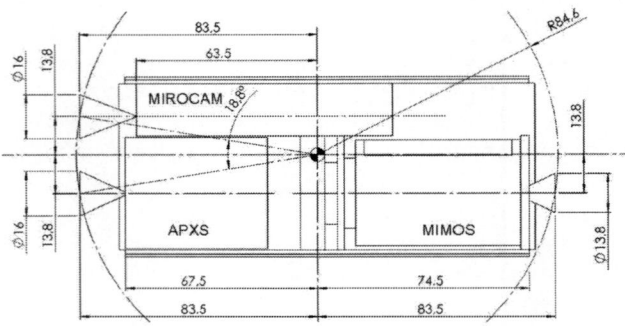

Figure 4.7: Accommodation of the GIPF sensor heads in the Nanokhod payload cabin

The articulation drive output axis aligns with the rotation axis of the PLC, which is mounted eccentrically off on the payload cabin. The position of the axis was chosen such that the rotation of the cabin allows to super–position fields of view of the three instruments. As the instruments vary considerably in size but also in focal lengths and aperture angles, the rotation around the PLC axis compensates for the differences between the integrated instruments. Figure 4.7 shows a cross section through the payload instruments with their field of views and the rotational axis. With this accommodation the principal measurement axis of the camera and of MIMOS coincide with each other when rotating the PLC by 180°. The measurement axis of APXS is inclined by 18.8° with respect to the other axis, when positioning the APXS field of view on the same sample, figure 4.8. The deviation of the axis has only negligible influence on the quality of the APXS spectroscopic measurements. This configuration allows the examination of a single sample with all three instruments by simply rotating the payload cabin. The principle provides a minimum effort of control and actuation and thus allows for reliable autonomous operation.

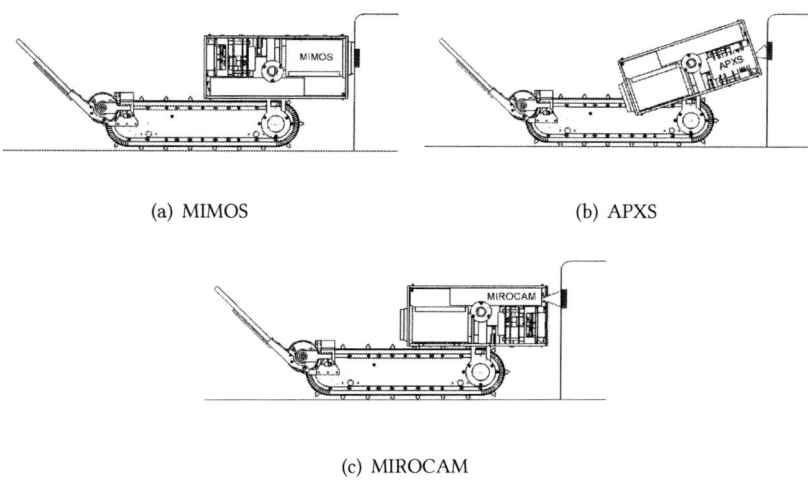

(a) MIMOS   (b) APXS

(c) MIROCAM

Figure 4.8: Super positioning of the GIPF field of views by a simple rotation of the Nanokhod payload cabin

4.5 Design of the rover drive system

Four drive units are used in the rover system. Two of which are used for the locomotion of the tracks and allow the rover to turn or rotate on spot, by driving only one LU or driving one forward and the other one backward. The two other drives articulate the payload cabin and provide two degrees of freedom for the positioning of the instrument heads on the sample. One of the articulation drives is situated in the front of the left LU and controls the attitude of the lever arm. The second drive is inside the PLC rotating it around its axis. Due to cost and reliability reasons, all four drive units used in the rover use a similar design and components.

4.5.1 Load requirements for the drive unit

The drive units are designed to meet the terrainability requirements of the BepiColombo mission. They have to provide enough torque to drive up a 20° slope on a regolith surface as well as to overcome a step height of 10 cm. The maximum required velocity of the rover is 5 m/h. Analyses show that the highest torque is needed for the requirement to recover from the situation when the rover is overturned. However the sealing on the LUs causes friction depending from the contact pressure of the sealing spring, which depends from the temperature condition. With the worst case friction on the sealing the maximum torque required for driving exceeds the torque necessary for the lever drives. The recovery needs enough torque of the PLC articulation drives to lift both the tracks and the tether unit above itself. The torque needs for the different required situations are listed in table 4.1.

Other aspects which are important for the MRP system are the reduction of mass and volume, whilst avoiding single point failures as well as achieving simplicity, robustness, high precision and a high efficiency of the drive unit.

4.5.2 Drive unit trade–off

The drive unit for MRP was completely redesigned with a new motor and gear stages integrated with each other into the rover system. For the previous Nanokhod model of the Micro–RoSA project the requirements for the drive system were more relaxed with Mars as mission target. The thermal requirements were less severe and in addition the planet provides an atmosphere which enables for the use of brushed DC–motors. Brushed DC–motors can not be operated

Parameter		Mercury	Earth	Notes
Gravity		2.78 m s^{-1}	9.8 m s^{-1}	
Locomotion	Slope	400 mNm	1400 mNm	20° slope
	Seal Friction	500 mNm	500 mNm	PTFE seal with 0.2 Nmm^{-1} contact pressure
	Total	900 mNm	1900 mNm	
Locomotion velocity		5 m/h	5 m/h	From MRP requirements document [24]
Obstacle climbing		450 mNm	1600 mNm	Worst case force for PLC on ground and lifting of LUs for Earth gravity conditions
Pointing accuracy		1°	1°	

Table 4.1: MRP expected load cases for the drive layout

continuously in vacuum, as the dissipated heat from the armature cannot be removed without an atmosphere providing for conductive and convective cooling. Contacting brushes also suffer from cold–welding effects between themselves and the armature. Thus the Micro–RoSA DC–Motor solution was not an option for the Mercury design, despite its advantages: With the brushes the motor has an inherent commutation, allowing for a reduced complexity of the drive. Compared to the stepper motor the no–load and the nominal current are lower which increases the efficiency of the motor and thus favours a DC brushed motor also for power limited applications.

A brushless DC motor is a possible alternative. It has some advantages over brushed DC motors including higher efficiency and reliability, reduced noise and longer lifetime due to electric commutation. Ionising sparks from the commutator are eliminated and electromagnetic interference is reduced. With the electromagnets located around the perimeter instead on the armature the thermal performance is more effective with direct conduction to the motor casing eliminating the need for air flow or atmosphere inside the motor for cooling. Without the brushes there is much less abrasion inside the motor internals and they can be entirely coated providing an additional sealing from dirt. A drawback of the brushless DC motor is the need for a more complex control electronics. To regulate the rotor commutation, the controller also needs some kind of encoder to determine the rotor orientation and position.

However, at the time of development there was no brushless DC motor available that provided the required torque and size. Finally a stepper motor was chosen for the MRP design. The stepper motor combines the advantages of not having a commutator and it also requires a less complex drive electronics. A disadvantage of the stepper motor though is that it has a generally lower torque and, adding to this, it may generate vibrations with the discrete steps of the rotor which can cause the motor to loose torque. Additionally to the limited alternatives the chosen stepper motor AM1020 has the advantage that it was already tested during the previous project RTPE for the required environmental conditions. The test confirmed that the motor works reliably down to $-180°$C [26].

For the former Nanokhod models the motor is attached to a planetary gearhead and then combined with a worm gear, which drives the sprocket wheels of the locomotion units or the PLC actuators. However, previous models have shown that the worm gear is susceptible to catastrophic failures and is highly inefficient. As this is a potential single point failure mode the drive design for Mercury aimed for a replacement of this gear type.

The gear stage is thus replaced with a Harmonic Drive unit, which provides a higher precision and overall a more robust drive system. In order to accommodate an angle of $90°$ between the rotational axis of the motor and the drive output, a crown gear is integrated as an additional stage. This configuration allows a compact drive unit which can be accommodated both in the locomotion units as well as in the payload cabin. Different Harmonic Drive gear combinations – with different gear ratios, different motor types and the integration of a micro Harmonic Drive (see appendix B) – were considered and traded against each other. The selected gear configuration is shown in table 4.2.

The output torque of all four drive units is 3.2 Nm and thus below the ratcheting torque of the Harmonic Drive of 5.2 Nm. The output speed of the drive unit is limited to ~ 3 m/h. This is due to the input rotational speed of the planetary gearhead which is limited to 5000 rpm for industrial use. The lifetime of the rover designed for the BepiColombo baseline mission is with two weeks much more limited than an industrial application, however the input speed was used as a basis for the drive layout. This approach was chosen as no reliable number is given for how much faster the input speed can be for the present conditions considering the short mission duration. This value has to be determined in the

	AM1020 motor	Planetary gearhead	Crown gear	HFUC 5–100	Track wheel/ Track exit
Speed	5000 rpm	78.13 rpm	32.17 rpm	0.32 rpm	3.03 m/h
Translation ratio	–	64:1	34:14	100:1	
Efficiency	0.69	0.7	0.85	0.7	
Min. torque [mNm]	0.7	31.36	64.74	3236.8	

Table 4.2: Transmission of the rover drive system

framework of a future project as part of the drive qualification. Furthermore, the use of industrial specifications for this project adds some additional margin to the lifetime of the drive design. The rover velocity was a result of a trade–off (appendix B) performed between the motors which matched the size and torque requirements and were available at the time of the design. The rover speed is below the required value of 5 m/h, but was decided to be acceptable, as it was the optimum solution at the time of the drive layout. It still includes the option to drive the rover with the required speed after careful testing of the gear unit for the required conditions and lifetime. However due to the limitation of the rover range to the length of the tether wire, 3 m/h are absolutely sufficient to travel the required 50 m during the two weeks of the baseline mission.

4.5.3 Setup of the MRP drive units

All drive units consist of a stepper motor, a planetary gearhead, a crown gear and a Harmonic Drive [27] shown in table 4.2.

The functional principle of a Harmonic Drive is shown in figure 4.9. A Harmonic Drive consists of three main parts: A circular spline, a flexspline and a wave generator. The wave generator is fabricated with a slightly elliptical shape and is enclosed by a small flexible bearing. The ball bearing is enclosed by the flexspline, a thin metal cup with teeth on its outer surface. The mating part to the flexspline is the circular spline, a rigid steel ring. The circular spline has on its inside surface two teeth more than the flexspline. The flexspline is deflected by the wave generator into an elliptical shape causing the flexspline teeth to engage with those of the circular spline at the major axis of the wave generator ellipse and with the teeth completely disengaged across the minor axis of the ellipse. The different number of teeth on the flexspline and the circular spline result into the transmission ratio of the drive.

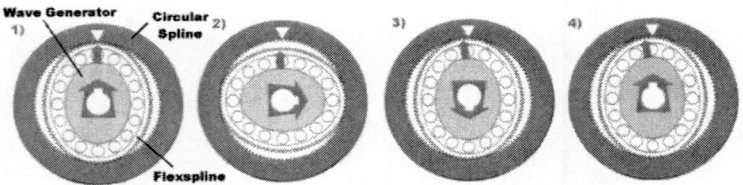

Figure 4.9: Principle of a Harmonic Drive [28]

The Harmonic Drive has inherently no backlash. The drive system backlash stems therefore only from the backlash of the planetary gearhead and the crown gear, reduced by the transmission ratio of the Harmonic Drive unit. This drive system provides high accuracy with a play of less than 5' [29]. This provides high precision for the positioning of the scientific payload instruments on the same sample.

The stepper motor is connected via the planetary gearhead to the pinion of the crown gear. The crown gear itself is attached to the wave generator of the Harmonic drive. The flexspline is rigidly mounted to the side wall of the locomotion units. The output of the drive is the circular spline that is connected to the sprocket wheel which is driving the track foil.

For the articulation drives the output of the Harmonic Drive is the flexspline and the circular spline is rigidly mounted to the wall of either the locomotion unit or the payload cabin. The flexsplines are attached to either ends of the left PLC lever arm. To ensure the position control of the payload cabin, a rotational encoder (see section 4.9) is integrated in the articulation drives. It engages with a thin key–shaft that is mounted to the bottom of the flexspline cup and leads through the inside of the Harmonic drive and the crown gear into the encoder.

Due to the different space constraints inside the payload cabin and the loco-motion unit, the solution for the crown gear attachment to the Harmonic Drive differs for the two articulation drives. In the locomotion unit, the mating surface of the crown gear points away from the Harmonic Drive as for the locomotion drive's setup. In the payload cabin the mating surface points towards the Harmonic Drive and the crown gear pinion is located in between the Harmonic Drive and the crown gear. Figure 4.10 shows the CAD model and the real component of the PLC Drive.

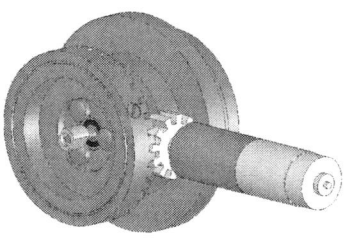

(a) CAD model of the PLC Drive   (b) Harmonic Drive Unit with stepper motor

Figure 4.10: Model and picture of the Harmonic Drive unit, the crown gear and the stepper motor with the planetary gearhead

The sprockets at the front ends of each locomotion unit, support and guide the track foil and are pivot mounted with bearings. For the left locomotion unit, the sprocket is mounted on the articulation drive with a 4–point bearing.

In order to seal the output of the articulation drives, a thin, customised O–ring is fitted between the circular spline and the surface of the PLC or LU.

4.5.4 Lubrication

Due to the constraints of the cold target temperatures, lubrication is only possible with dry lubricants. All contacting surfaces inside the Harmonic Drive and of the crown gear are lubricated with Microseal 200–1, which is a modified embedded tungsten disulphide (WS_2) lubrication, known under the registered trade name Dicronite®. The lubricant is conform to AMS 2526, which has passed the relevant ESA outgassing tests. The lubricant is embedded into the surface of the part at a thickness of about 5 to 10 micron. It is applied by a special cold spraying process at room temperature, which incorporates the lubricant into the material. Although a thin film of WS_2 remains on the surface of the material, it is not necessary for lubrication and can be removed without any negative effect. This ensures that the dimensions of the component will remain the same. The friction coefficient of two contacting parts is reduced to 0.03 [30].

For the MRP model, the bush bearings of the stepper motor are not lubricates, while the planetary gearhead is lubricated with a MoS_2-bath. This lubrication procedure for the motor + gearhead is based on the successfully tested configuration of the RTPE project (see section 4.5.2).

However this configuration allows only for limited radial loads due to the not lubricated bearing and needs to be reworked for the flight model. For the flight model the bearing of the motor needs to be customised and dry lubricated by the motor supplier before the assembly of the motor. The financial limitation of the project did not allow for such a customised motor development.

During the drive unit development the properties of the two mentioned types of dry lubrication were traded against each other for this application. It was considered to implement Dicronite®for the whole drive unit, due to the better suitability to humidity. This would allow for easier conditions during the assembly and testing phase. However this was discarded during this development step due to the missing experience combined with a higher effort.

As part of the next development step the properties of the two dry lubrication types should be thoroughly tested, specifically concerning their friction coefficients for the considered temperature range and their handling properties during application and testing.

4.5.5 Torque tests on the drive system

Tests on the drives system have clearly demonstrated that it is capable of producing sufficient torque for the application. However, during the initial functional testing the drive occasionally stalled whilst running. Investigations indicated that a number of factors are causing the torque margin of the stepper motor to be significantly reduced, which ultimately could cause the motor to miss a step. Due to the mass inertia of the rotor, the motor is unable to continue at operational speed. The rotor needs a controlled acceleration ramp; this is why the missed step invariably leads to the motor stopping.

When the stalling of the drive first occurred during the rover functional tests, several examinations were carried out with the following results [31]:

- Software/electrical drive problems were discounted after tests.
- Stalling position suggests no single tooth of the crown gear is responsible for the stopping of the motor.
- Load is a factor in the frequency of occurrence.
- The problem occurs more often in one direction of the output shaft (but

opposite for the two PLC drives). The crown gear is mounted in different orientation for PLC–drive (teeth towards HD) and lever–drive (teeth away from HD). This means that the crown gear orientation correlates with a the higher probability of errors. This can be an indication of a slight machining error.

Motor tests on a spare motor showed that the method of clamping the motor could cause the motor to stop, but also that it is more than capable to produce the required torque. The re–testing of the drives with an improved motor clamping method showed some improvement. Another improvement was achieved by increasing the distance between pinion and the crown gear to a larger amount than required by the setup tolerances. The bigger distance reduces the radial loads on the planetary gear axis which reduces the generation of torque. Still the problem could not be completely solved and the stalling continued to occur occasionally. However the problem appeared less frequently the longer the drives were operated. Some running–in effect seemed to improve the situation. This correlates with the friction characteristics of the Dicronite lubricant over the number of revolutions [32]: The friction is approximately 3 to 4 times higher for the first revolutions, after a running–in time it drops to a lower friction coefficient of 0.03, before suddenly rising excessively at the end of its lifetime (>10000 revolutions).

The mounting tolerances over the whole drive system interfaces are another probable factor leading to the reduction of torque margin. The Harmonic Drive unit comprises already nine interfaces. Also the dry lubricant may increase the stiction forces of the gear teeth faces. All effects are exacerbated by the small scale of the components relative to the torque levels used.

4.5.5.1 MRP drive unit test with stepper motor

Although the problem seemed to disappear through the running–in of the system the drive was carefully examined. A special test campaign was conducted at the Harmonic Drive AG (HDAG) in order to investigate the problem and verify the drive system and its torque margin [33]. The campaign consisted of two tests. The objective of the first test was to assess the MRP drive system (consisting of the planetary gearhead, the crown gear and the Harmonic Drive unit (HFUC–5–100) driven by the stepper motor AM1020), when mounted in a rigid environment on the test bench. The test setup provided assessments of torque variations in the stepper motor as well as acceleration measurements.

The objective of the second test was to verify the torque loads in the Harmonic Drive during the revolutions of the gear. This drive setup was performed with a DC motor in order to monitor the current which is directly related to the required torques.

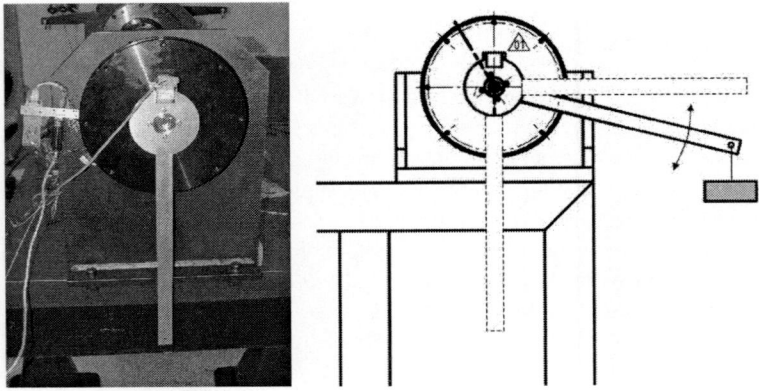

Figure 4.11: Test rig for drives test with test setup of the stepper motor test

The test–setup for the stepper motor tests is shown in figure 4.11 and described in more detail in [34]. The schematic of the stepper motor test is shown in figure 4.12.

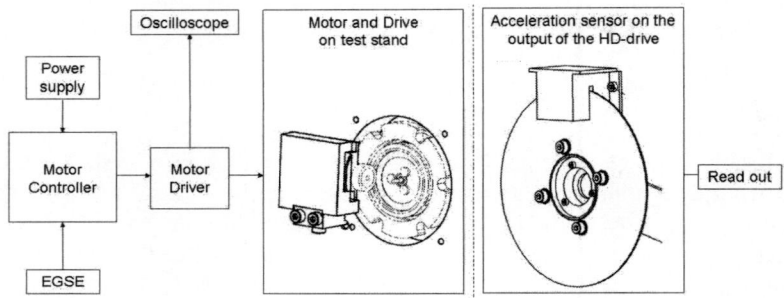

Figure 4.12: Schematics of the stepper motor test setup

For the test a 50 cm arm was attached to the Harmonic Drive flexspline output

of the LU lever drive to apply load torque to the drive. The arm had a bucket at the other end to add weight for adjusting the torque of the test setup. The arm started from the low vertical position and moved to the horizontal position. Without the bucket and any additional load, the maximum torque to move the arm in the horizontal position was 0.35 Nm. By adding weight in steps of 60 g the torque applied to the drive is raised to maximum 2.7 Nm for the horizontal arm position. For each load step the arm is driven counter–clockwise from the vertical to the horizontal position and back down again. The test is repeated with three different step rates: the nominal step rate of 1528 steps/s, the step rate of 934 steps/s and the slowest step rate of 104 steps/s.

The armature of the stepper motor is driven by switching in between the phases of the stator. For each step the voltage applied to the coil of one phase is switched off and is applied to the coil of the next phase, causing the armature to move on. After the switching between two phases the current needs a so called *rise time* to build up in the coil.

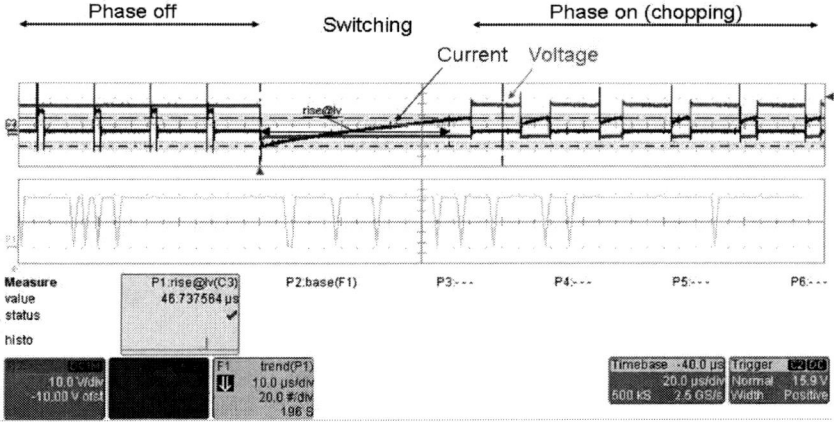

Figure 4.13: Oscilloscope display showing the rise time after switching to the next motor phase

Figure 4.13 shows the current and the voltage measured when switching a phase from off to on in between two steps. The motor is driven with a chopper drive circuit, such that short rise times can be seen also when the phase is on. For the chopper drive a high–voltage is applied to provide rapid changes in current when switching a phase on or off. This helps to approximate the current waveform to the ideal square current waveform. The high–voltage has to be chopped, to avoid overheating the motor. For the chopping, the voltage is switched off, when the current exceeds an upper limit and back on again, when the current falls below a lower limit. The chopping thus causes the current to show a rising level, however due to the set lower and upper limits the ramps and the resulting rise time are much shorter than for the switching between the phases, which is more difficult to measure and thus the measurement focusses on the rise time between the switching.

The rise time depends on the inductance of the motor coil. The inductance itself is dependent on the air gap between the stator and the armature. With a load applied to the motor the stator and armature do not superimpose completely and the air gap increases, which causes the inductance to decrease. With the lower inductance the current needs less time to build up and the measured rise time is shorter. Thus the measured rise time correlates to the torque load applied to the motor [35], [36].

An oscilloscope was utilised in the test setup to measure the change in inductance of the motor windings when load is applied to the stepper motor. The oscilloscope measured the voltage as well as the current (voltage measured across a $1\,\Omega$ resistance).

The test results are described in the following paragraphs. During the tests no jerking of the drive system occurred, nevertheless the measurements of the first test confirm the suspicions for the cause of the stalling. The measurement results for different applied loads and for the different step rates are shown in the figures 4.14 to 4.16. The graphs present the rise time for the reversion of polarity in between two steps applied over the time to lift the arm from vertical to horizontal. The general trend of the measurement is shown in the graph with a green curve derived from an average of 40 measured values.

Figure 4.14, with the nominal step rate, shows a clear trend of a decreasing rise time for the arm movement. The trend for all three loads shows an increasing torque load equivalent to the decreasing rise time, as the torque arm moves into the horizontal position. Several torque peaks can be seen in all graphs with the spikes of a shorter rise time. This trend becomes more obvious with a higher load added to the torque arm. These graphs explain the higher occurrence of the problem with a heavier load applied to the drive, as the torque peaks add up to the general load and thus can lead to a torque decrease which is larger than the motor margin allows for.

For the nominal step rate of 1528 steps/s the peaks seem to be independent

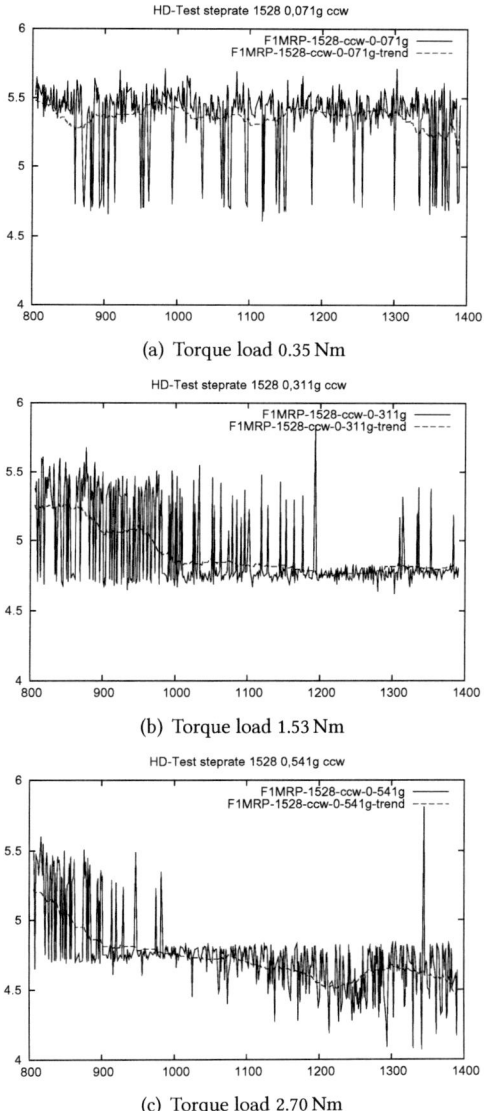

(a) Torque load 0.35 Nm

(b) Torque load 1.53 Nm

(c) Torque load 2.70 Nm

Figure 4.14: Rise time distribution over a 90° counter-clockwise output movement for different loads and the nominal step rate of 1528 steps/s

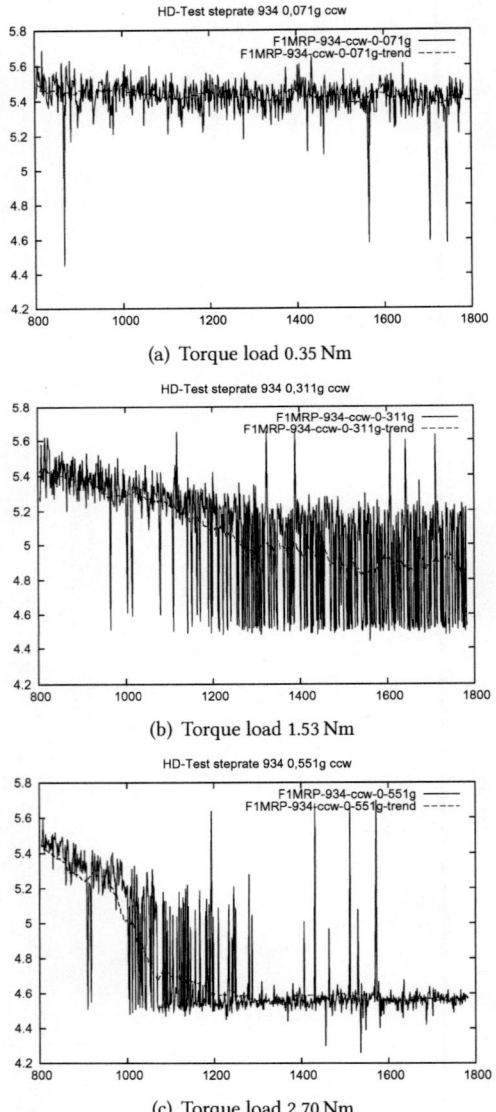

(a) Torque load 0.35 Nm

(b) Torque load 1.53 Nm

(c) Torque load 2.70 Nm

Figure 4.15: Rise time distribution over a 90° counter–clockwise output movement for different loads and the reduced step rate of 934 steps/s

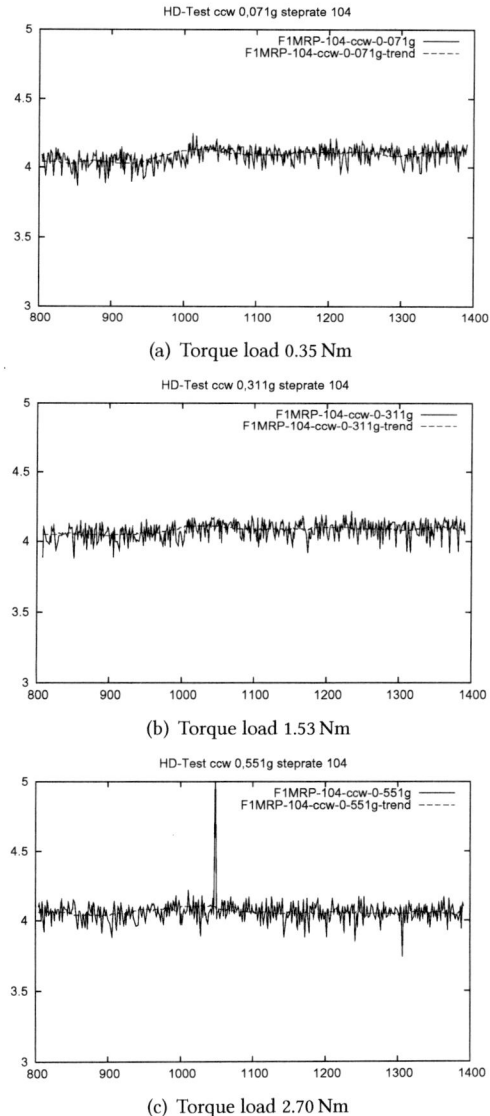

(a) Torque load 0.35 Nm

(b) Torque load 1.53 Nm

(c) Torque load 2.70 Nm

Figure 4.16: Rise time distribution for a range of ~ 1,4° for each the vertical, 45° and horizontal load arm positions for different loads and the slowest step rate of 104 steps/sec

from the torque load. However for the reduced step rate of 934 steps/s the overall number of peaks clearly decreases. A high number of peaks is only shown for an increasing torque, for the test with a medium load applied, figure 4.15(b). These results suggest that the slower step rate generates less vibration in the stepper motor.

For the slowest steprate of 104 steps/s (figure 4.16) not the full angle range of the load arm is recorded due to limitations of the recording memory. Instead, a measurement is recorded for a range of ~ 1,4° for each the vertical, 45° and horizontal load arm positions. All three measurements per load case are shown together in one graph with a mark separating the different angle positions.

While the drive behaves as presumed for the nominal and the medium speed, the graphs for the slowest step rate do not show a decreasing rise time for an increasing load. On the contrary for the first two load cases (figures 4.16(a) and 4.16(b)) the rise time even increases slightly for the lifting movement of the arm, however with a much smaller delta than the previously observed rise time changes. It has to be noted that the measurement of the rise time is not an explicit evidence for the increase of the load torque.

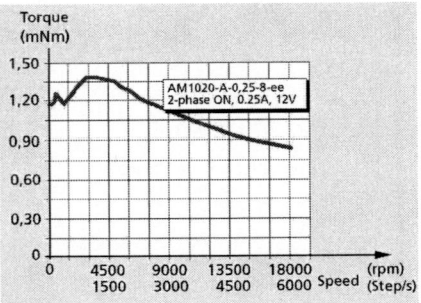

Figure 4.17: Torque–speed–diagram for the current mode of the AM1020 stepper motor – figure from Arsape [37]

Considering the torque–speed–diagram of the AM1020 stepper motor (figure 4.17 [37], it becomes obvious that the first two test steprates are well within the speed range of the maximum torque. However the steprate of 104 steps/s is in a range of the diagram, which shows several dips. These dips for slow step rates result from resonances of the single step response. The oscillation of the armature causes changes in the inductance of the motor and thus the rise time.

The vibration effects are highly dependent on the damping of the system, for example depending on the increasing load torque, when the arm moves up.

This is why the rise time increases, though only marginally, with the rising torque of the lifting arm for the first two loads applied. For the maximum weight (figure 4.16(c)) the higher torque load compensates the increasing rise time effect for the slow steprate.

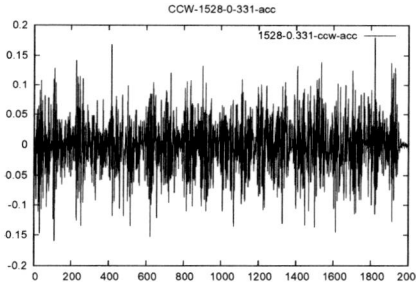

Figure 4.18: Acceleration on the output measured by the Ferraris sensor for a test with the nominal step rate and a max. torque load of 1.53 Nm

Figure 4.18 illustrates a typical measurement of the acceleration sensor. There is no evidence of any influence of the different gear stages as the data is too noisy at first sight. In order to verify if the data is a combination of sinusoidal basis functions, which identify specific influences, a Fast Fourier Transformation (FFT) analysis was carried out on the data of the Ferraris acceleration sensor.

The comparison of the FFT analysis for the nominal step rate with two different loads are shown in figure 4.19 and show the noisiness of the acceleration data. However, for both torque loads the same two peaks can be seen in the range of ~ 22 Hz and in the range of ~ 47 Hz. The amplitude of these frequencies varies with the torque load.

Comparing the data of the FFTs of the acceleration output for the different step rates (shown in figure 4.20 (step rate 934) and 4.21 (step rate 104)) it is obvious that the system shows the same frequencies, also for the slower step rates. The fact that the frequency is the same for all motor speeds implies that this is the eigenfrequency of the gear unit setup. The amplitude of the frequencies is the only parameter that varies with the motor speed, but also with the applied torque load.

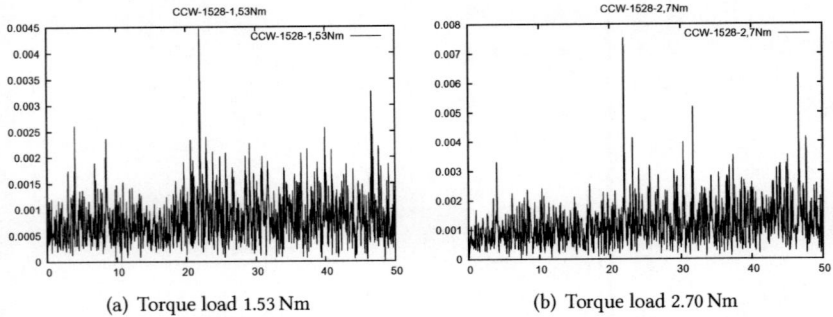

(a) Torque load 1.53 Nm   (b) Torque load 2.70 Nm

Figure 4.19: FFT analysis for stepper–drive tests with different applied torque loads and the nominal step rate of 1528 steps/sec

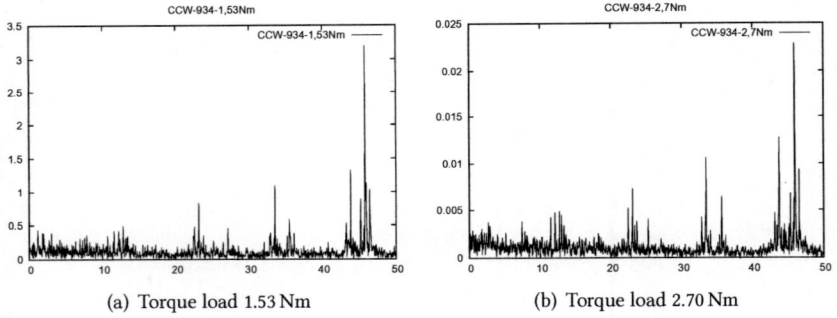

(a) Torque load 1.53 Nm   (b) Torque load 2.70 Nm

Figure 4.20: FFT analysis for stepper–drive tests with different applied torques loads and the reduced step rate of 934 steps/sec

The medium step rate (figure 4.20) shows the most defined eigenfrequency with several harmonics. It can be derived that this step rate provides the smoothest running of the system and that there are only little disturbance frequencies. However a specific influence of one of the gear stages can not be detected from any of the acceleration data graphs. A comparison with the frequencies of the different gear stages in table 4.3 does not indicate that the shown peaks are driven by an irregularity of one of the gear stages. However the impact of the gear–stages can not be completely disregarded as it might be covered by other effects.

For the next development step it is essential to plan for tests of the separate gear stages, before assembling the unit and integrating it into the rover. Only proper analyses of the drive unit will provide a reliable drive system for the flight model of the rover.

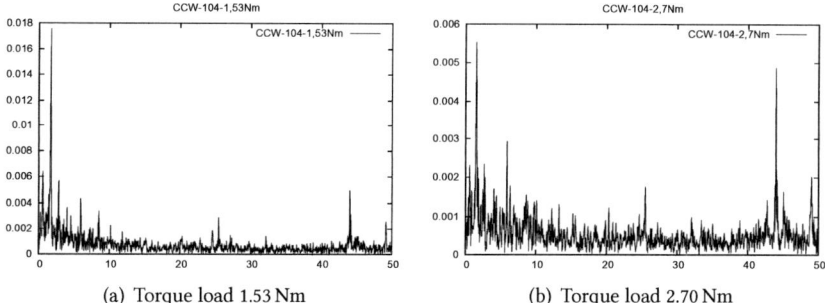

        (a) Torque load 1.53 Nm   (b) Torque load 2.70 Nm

Figure 4.21: FFT analysis for stepper–drive tests with different applied torques loads and the slowest step rate of 104 steps/sec

Gear stages	Step rate 1528 steps/s		Step rate 934 steps/s		Step rate 104 steps/s	
	Angular frequency [rad/s]	Frequency [Hz]	Angular frequency [rad/s]	Frequency [Hz]	Angular frequency [rad/s]	Frequency [Hz]
Harmonic Drive	0.49	0.08	0.30	0.05	0.03	0.005
Crown gear	1.19	0.19	0.73	0.12	0.08	0.013
Planetary Gearhead	76.34	12.15	46.70	7.43	5.20	0.828

Table 4.3: Frequencies of the drive stages for the different motor speeds

4.5.5.2 MRP drive unit test with DC–motor

The objective of the second test is to verify the torque loads specifically for the Harmonic Drive gear stage. For this test the original motor, as well as the gear staged before the Harmonic Drive, are replaced with a DC motor (PMA-11-100-01-ES00ML), figure 4.22. The application of torque loads is done in the same way as for the stepper motor tests: The drive is mounted onto a rigid test bench and a load arm is applied to the output of the Harmonic Drive. The arm is driven from the vertical to the horizontal position with different loads added on the other end of the arm, providing an increasing torque for the lifting of the arm.

Figure 4.22: Test setup for second Drive test with a DC motor driving the Harmonic Drive

During the test, the torque that is theoretically needed to back–drive the gear was measured. The objective of the test was to verify, if the Harmonic Drive produces torque peaks which were observed in the stepper motor setup. The input speed on the Harmonic Drive input side is 26 rpm.

The graphs in figure 4.23 show the back–drive torque for three different load cases with an increasing weight added to the arm from left to right. As expected, the trend (derived from an average over 40 values) of the back–drive torque is deflected from the horizontal axis for an increasing load of the arm travelling from vertical to the horizontal position. This effect becomes more obvious for increasing loads added to the load arm. Torque peaks as for the stepper motor tests can not be detected. However the measured data is relatively noisy, resulting from the sensitive drive control.

This separate test of the Harmonic Drive unit does not indicate any irregular running of the Harmonic Drive. The eigenfrequencies derived from the FFT of

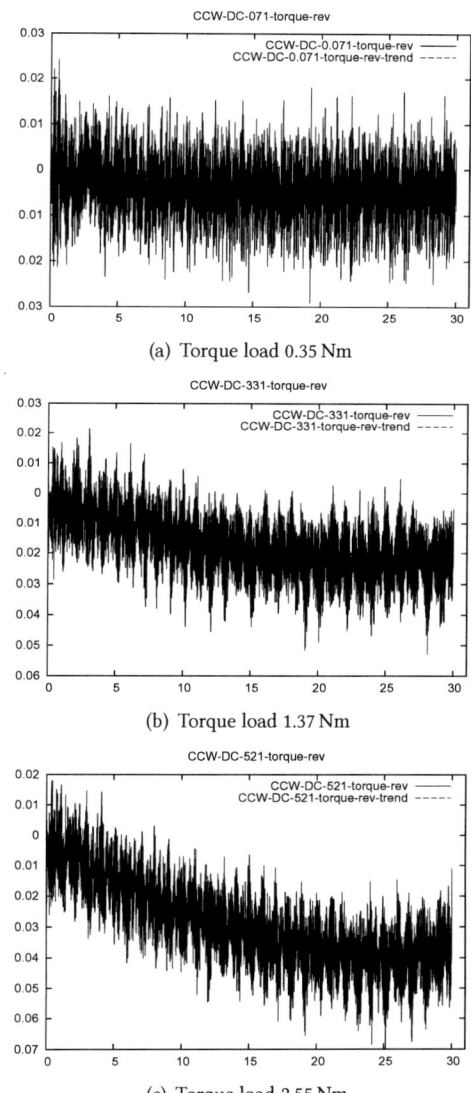

(a) Torque load 0.35 Nm

(b) Torque load 1.37 Nm

(c) Torque load 2.55 Nm

Figure 4.23: Back–drive torque over a 90° counter–clockwise output movement for different loads

the acceleration data and shown in figure 4.24 do not overlay with the frequency of the Drive unit rotation, given in table 4.4. Due to the variation of the test setup and the disassembly of the drive units the frequencies for the DC test differ from the system frequencies of the stepper motor setup. The impact of the different applied torque loads becomes again obvious in the variation of the frequency amplitudes. As suggested before, further tests of the single drive stages should be conducted, including the repetition of this test with varying motor speeds, in order to reveal any correlations.

	DC	
Gear stages	Angular frequency [rad/s]	Frequency [Hz]
PMA	0.43	0.07
Harmonic Drive	43.35	6.90

Table 4.4: Frequencies of the Harmonic Drive for the DC motor test

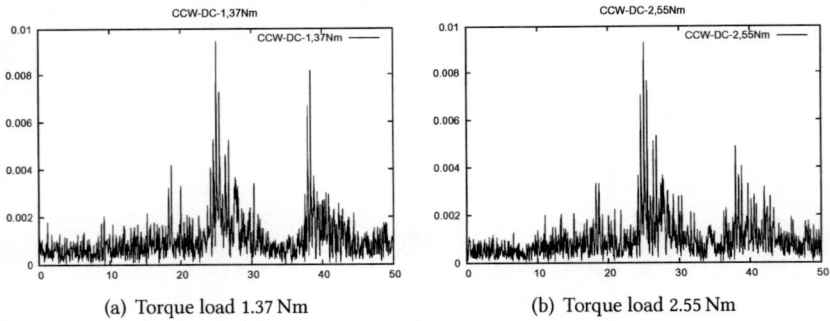

(a) Torque load 1.37 Nm   (b) Torque load 2.55 Nm

Figure 4.24: FFT analysis of the acceleration data for PMA–drive tests with different applied torque loads

4.5.5.3 Thermal test of an alternative actuator – the brushless EC10

In the course of the project an alternative actuator was introduced to the market. The brushless EC10 is of similar size as the currently used motor AM1020 and it is able to provide a higher torque as well as a higher output speed of the rover. In the framework of the drive system analyses a cold test of this motor was conducted, to confirm the functionality of the brushless motor for the temperature range down to $-180°$C.

For the test a sensorless brushless DC (BLDC) motor and a brushless DC motor with Hall sensors, both of the type EC10, were tested [38]. The motors were controlled with a digital EC controller for a sensorless BLDC, which measures the back EMF (Electro–Magnetic Force) to control the motor. The Hall sensor signals are thus not necessary for the operation of the motor. However an oscilloscope was used to monitor the functioning of the Hall sensors during the test, to see whether it is capable to work within the required temperature range. In order to cool the motor to $-180°$C it was attached to a cooling loop fed with liquid nitrogen, see figure 4.25.

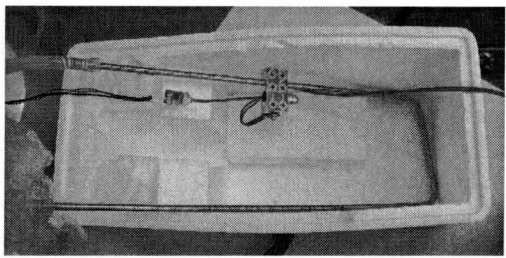

Figure 4.25: Test setup for deep temperature test of the brushless DC motors

The first test with the sensorless motor showed that the motor works down to $-180°$C. Initial problems to switch the motor on and off at low temperatures were solved by repeating the degreasing procedure of the motor. After that, the motor worked reliably over the whole temperature range. The test with the DC motor with Hall–sensor also worked reliably down to $-180°$C as expected, however the Hall sensor signals failed. The signals started to disappear at a temperature of approx. $-80°$C. One explanation for the malfunctioning of the Hall sensors is that the electronics cannot measure the Hall Effect voltage in the deep temperature range as the characteristics of the effect changes.

To use the brushless DC motor for the rover drive design the encoder unit of the motor thus has to be reviewed, because the DC driving method using the measurement of back EMF is not advisable for this application. The back EMF is very low at low speeds, which makes the motor control very difficult.

In order to be able to retain the brushless DC motor as a promising alternative actuator the following steps could solve the encoder problems: One is to revise the current electronics and consider the Hall signal detection which may allow the Hall sensor control of the motors down to −180°C, however it is doubtful that the Hall Effect can be used over the considered temperature range and it might require a completely new design of the Hall sensor and its electronics. Another way is the replacement of the Hall sensor with another encoder type which works reliably down to −180°C. Nevertheless the tests have proven that the motor works repeatable down to the required temperature range. This offers the possibility to redesign the drive unit and to implement a motor which allows for a bigger torque margin, resulting into a more reliable drive unit.

4.6 Design and aspects of the tether unit – Interface to the lander

The tether unit (TU) in the back of the rover provides the rigid link between the two locomotion units. The tether unit houses a PCB with the tether interface electronics and supports the two spools with the tether wires, that connect the rover with the lander.

In the new rover design the tether unit is moved upwards to provide more ground clearance and to avoid a bulldozing effect on a regolith surface, which was seen on previous test models. The lifting of the TU induces that the top/bottom symmetry for full operation was abandoned. The loss of symmetry was considered to be less critical as the mission control would take a very conservative operational approach, avoiding any risk which may lead to the tipping over of the rover in order to not damage it and endanger the mission. Additionally, it is still possible to turn the rover back over. After giving up the symmetry, the tether guides were also changed so that they point upwards at an angle of 45°. This provides a better protection of the tether wires when moving backwards and avoids that dust enters through the guides.

The tether wires consist each of 50m long HF (High Frequency)–stranded wires that are coiled onto two spools. The electrical connection from the spools to the rover is provided by spring pre–loaded gold plated slip rings that are sealed against dust. The two spools are mounted onto an axis using a set of ball bearings, which allows the free deployment of the wires, as the rover moves forward.

During the transfer of the rover to the planetary surface the tether guides are folded to the front of the rover and secured with a pre–tensioned polyethylene strap, e. g. high–strength Dyneema. For the deployment of the tether guides the strap is cut with a thermal knife, and a spring load moves the guides into position with an angle of 45° respective to the planets surface.

4.7 Design and aspects of the payload cabin lever arms

The payload cabin is held by two lever arms on each side connecting it to the locomotion units. Apart from the mechanical support of the payload cabin the lever arms fulfil two different purposes:

The lever arm on the left side is connected to the two articulation drive units, one situated in the front of the left locomotion unit and the other one in the payload cabin. The arm is used as a structural element for the articulation of the payload cabin.

The right lever arm is used for the electrical connection to the payload cabin, by providing the route for the power and data harness. In order to allow freedom of rotation of the payload cabin, the harness is coiled into a spiral at each end of the arm. This allows a rotation of the arm and the PLC of 370°. The upper spiral is integrated into the arm due to limited space inside the payload cabin. On the lower end of the lever, the harness spiral is integrated inside of the locomotion unit to minimise the restriction on the ground clearance of the rover.

4.8 The electrical design of the rover

The temperature range required for the BepiColombo mission varies between −180°C and +70°C. No off–the–shelf electrical components are qualified for this range. Due to that some of the used components, especially critical parts were tested down to −180°C in the framework of this project [39]. These tests also examined if parametric changes of the measured values occured. The parts chosen for the rover design were not necessarily the ones behaving as specified, but the components with reliable and predictive functionality. For the applied electrical design approach the following guidelines can be defined, which avoid or reduce deep temperature effects to a large extent:

- Critical signal processing should be implemented in digital technology rather than analogue, as the latter is more susceptible to parameter variations over the temperature range.

- If analogue measurements become necessary, they should be carried out supported by a reference value.
- Active components should be of the MOS–type (Metal Oxide Semiconductor), rather than the bipolar–type; the gain of a bipolar transistor for example tends to zero for temperatures near $-190°$C.
- The design should not rely on high accuracy RC timing constants.

Alongside the thermal requirements, the electrical design is furthermore affected by the radiation environment of a mission to Mercury. The maximum radiation level for the mission is expected to be 22 krad(Si). Although no radiation resistant electrical components were used for the current engineering model, this was still considered during the electrical design. All components chosen in the current design do have a radiation resistant equivalent or a technology path to a rad–hard version. In this manner, a radiation hard electrical design would not require a major redesign.

The rover power is provided by the lander batteries. With a given mission duration, the consumed energy relates directly to the battery capacity and thus to the battery mass. Thus, the energy consumption is a significant design driver of the rover system. During the mission, the rover has four different levels of energy consumption, based on the following activities:

- Rover locomotion.
- Localisation of the rover position supported by the lander.
- Instrument deployment by positioning the PLC.
- Scientific measurements (divided into the three cases for each instrument).

Additional to these four activity phases the rover will have periods, when it is not active, but waiting for communication with Earth for further operational instructions. For a profile as defined for the baseline mission, the rover will only spend ~ 14 hours in the 14 days performing locomotion. The localisation of the rover and the PLC movement for instrument deployment are relatively insignificant in duration. In comparison, the times required to perform reasonable measurements with the spectrometers are expected to be approximately 1 respectively 3 hours minimum per measurement for APXS and MIMOS – or in total greater than 114 hours during the baseline mission of 14 days. However the rover mode with the longest duration is the idle state, when the rover is awaiting input from Earth. It should also be noted that whereas the locomotion duration has a hard limit defined by the length of tether wire, the measurement quality is improved with longer measurement duration which may be requested by the scientists during operations. Based on these facts the energy consumption was minimised for the measurement phases and for the non–active periods of the rover. The energy consumption of the rover for a 14 day operation is shown in table 4.5.

Activity	Description	Duration		Energy consumed	
		Hours	% Total	Wh	% Total
Checkout	System test on lander	8	3%	13	4%
Deployment	Exit from lander, perform first measurement cycle	9	3%	17	6%
Movement	Move between two measurement sites	14	5%	76	26%
Measurement cycles	One measurement with each instrument	114	36%	190	64%
Idle	No activity	222	70%	0	0%

Table 4.5: Power consumption for the different operational modes

4.8.1 Electrical tether interface

External communication and power are passed to the rover from the rover interface situated on the lander via the two tether wires linking rover and lander. Each tether wire consists of 30 single Cu–LS high frequency strands with a strand thickness of 54 μm. The strands are enamel–insulated and the tether wire is silk woven. The usage of strand wire has the advantage of a very flexible connection between the lander and the rover. The wires can be stored on small freely rotating spools inside the rover and are deployed while the rover is moving over the planetary surface. The separation of the wire in many strands offers redundancy, as the breakage of a single strand will not cut off the complete connection. The stranded wires are furthermore designed to reduce the electrical losses caused by the proximity effect.

Power and communication signals are superimposed on two combined signals at the tether interface on the lander and then passed to the tether unit, which interfaces with the external connection to the internal rover bus. One of the tether

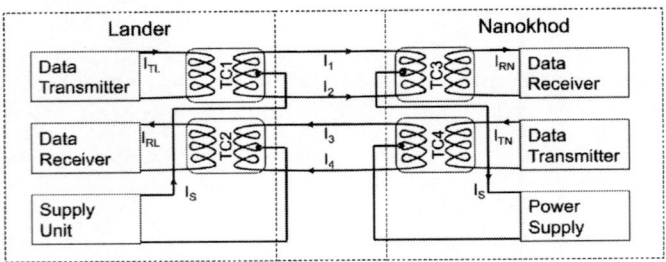

Figure 4.26: Schematics of the electrical tether interface

wires is used for the power supply towards the rover while the other one is closing the loop back to the lander. Each tether wire is again split into two electrical separated bundles in order to transfer the differential communication signals; this allows the control signals to be sent via one tether wire while the other one transmits the gathered telemetry data back to the lander. The differential signals are always coupled to one bundle by a transformer as shown in figure 4.26.

The losses of different methods of power transmission have been compared within the MRP–project [40]. In the former Nanokhod projects (MicroRoSA [9] and RTPE [10]) the connection to the lander was also realised with tether wires but using a high voltage transmission concept, based on the proportionality of the power losses to $\sim \frac{R}{V^2}$. This means that a voltage of 28 V DC was up–converted to 100 V DC at the tether interface on the lander side for the transmission across the tether in order to minimise the ohmic losses of the tether wire. Even though analyses during the RTPE study showed that the main losses were caused by the up and down–conversion, the baseline for RTPE was still kept with the high voltage DC transmission in order to meet the standard spacecraft practise.

For MRP one of the main requirements was the very limited power supply from the lander. In order to achieve those, the losses had to be reduced as far as possible. Within this project the alternative of having a low voltage transmission (28 V DC) was compared to the former high voltage DC transmission and additionally the high voltage AC transmission. It was found that the losses for both kinds of high voltage transmission for high and low power rover modes were higher than for the low voltage transmission. Avoiding up– and down–conversion also provides a mass and volume advantage. The only drawback of this option is the lack of galvanic isolation from rover to lander, which improves the EMC

(Electromagnetic Compatibility). The standard practise is to fully isolate the payload, if it is connected to a fully isolated DC–DC–linkage. However, for highly mass restricted missions the trend is to have only one central power conversion unit, in order to save mass, as e. g. on Beagle 2.

4.8.2 Internal bus

With the previous Nanokhod models one significant problem was always the routing of the cables. The cable routing from the locomotion unit into the payload cabin always got in conflict with the positioning of the PLC actuators. To reduce the number of wires a 9–wire bus system was implemented in the rover, which includes a two wire serial bus for communication. The serial bus uses the I^2C (Inter–Integrated Circuit) protocol for the exchange of the internal rover data.

The electrical system of the rover is divided into several nodes, as shown in figure 4.27, which are equipped with an I^2C interface and are acting as functional blocks.

4.8.3 Rover nodes

The rover's electrical system is split into Five nodes. Two of these nodes, the rover central controller node and the tether data unit node are the core system and always remain powered. The rover central controller is located inside the payload cabin and the tether data unit is in the right locomotion unit. The other three nodes are the motor controller nodes, which are only powered while they are in use. The status of the motor controller nodes is controlled by the 28 V line, which is at the same time the power supply of the motors. Two of the motor controller PCBs are located in the locomotion units and one in the payload cabin. The motor controller have an interface of up to two motor driver PCBs, which interface to the mechanical drive system.

All nodes collect housekeeping data from the sensors distributed in the rover and transmit it via the serial bus. The data is collected from the accelerometers, the contact switches and the rotational encoders of each of the PLC actuators (see also section 4.9).

4.8.4 Motor driver

The motor driver PCBs are designed as separate boards, so they can always be situated close to the actuator. This minimises the electromagnetic interference

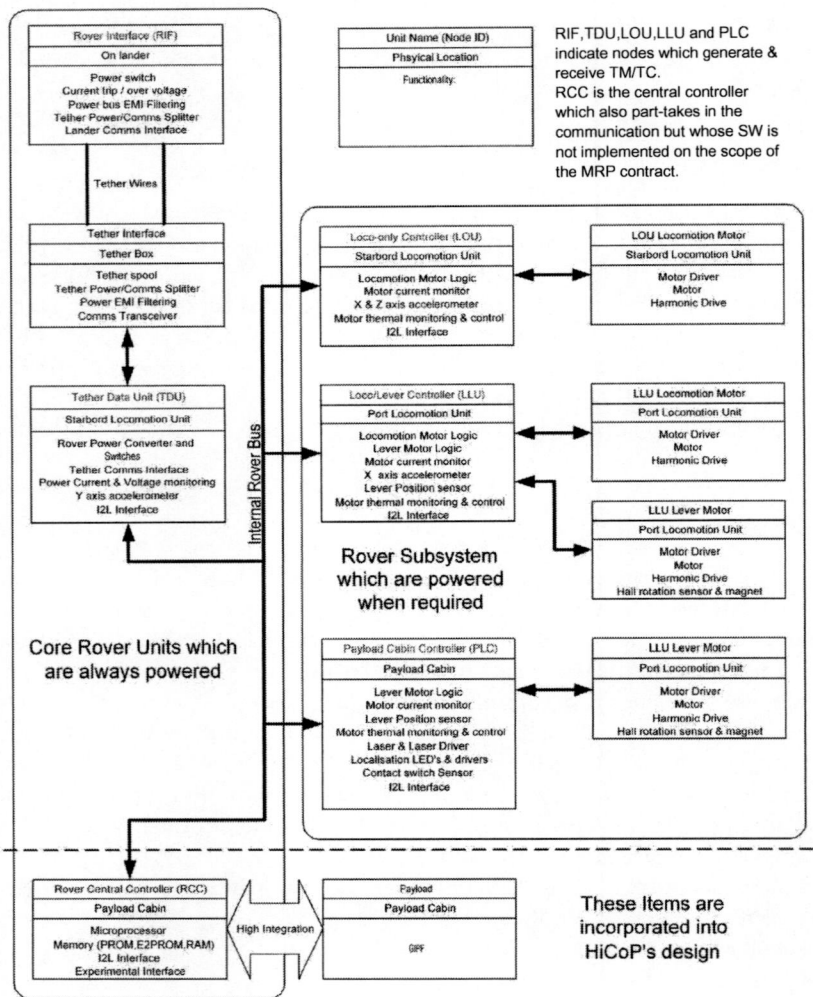

Figure 4.27: MRP Rover Internal Bus

between the board and the motor and provides at the same time the EMC separation to the main node. The driver PCBs activate the motors by current mode and with a constant speed of 5000 rpm. The driver PCBs of the LU are all the same, only in the payload cabin the driver had to be integrated onto the PLC–PCB due to limitations of space.

4.9 Rover sensors

The rover is equipped with various sensors needed for the control concept. With the data from these sensors, the rover subsystem together with the lander can be operated autonomously. The sensors provide information on the location, the orientation as well as on the inclination of the rover. Thus dangerous situations can be detected without a time consuming interaction with ground. Furthermore the sensors give the rover system some information about its near surroundings, so that it can detect either obstacles that need to be avoided or samples which are of scientific interest. The sensors also detect how the payload cabin is orientated and with the contact sensors if it is in contact with the surface or an obstacle. This allows for autonomous positioning of the scientific instruments.

For the MRP project the gathering of sensor data and housekeeping is done very extensively, in order to learn as much as possible from the breadboard and the tests carried out. Also during the space mission as much information as possible is collected, as this is the only data operational decisions can be based on. The housekeeping information collected by the Analogue–to–Digital Converter (ADC) are the currents and voltages of the different nodes. Also temperatures of critical parts are measured and saved in the housekeeping with integrated thermistors PT1000.

Further sensors which are used for the control concept in this design are described in the following sections.

4.9.1 Line laser

An integrated line laser in combination with the camera forms a simple system to model the surrounding environment within the rover processor. This allows a limited autonomous navigation without any interaction from the lander or the GSE side.

An algorithm calculates the environmental model by checking the deviation of the laser line projected onto surrounding obstacles as imaged by the MIROCAM.

In order to gain a sufficiently detailed model the MIROCAM takes a sequence of eight pictures, in between which the payload cabin is rotated by a small predefined angle. The resulting environmental model can then be used to detect obstacles or areas of interest for scientific sampling, their approximate shape and their position.

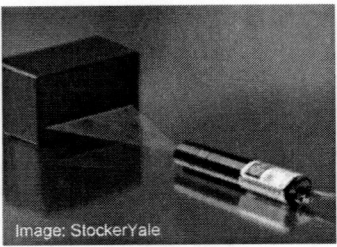 

Figure 4.28: Line Laser accommodated in the rover control partition of the Nanokhod payload cab

4.9.2 Inclinometer

Three orthogonal accelerometers are working together as an inclinometer. Two accelerometers are situated on the motor controller PCBs in the locomotion units and one is on an extra PCB, which is connected to the Tether Data Unit PCB. The devices used are the ADXL203 by Analog Devices, which showed superior performance during cold temperature tests.

4.9.3 Contact sensor

For the positioning of the MIMOS instrument a contact sensor is located on the payload cabin in front of the MIMOS instrument. It consists of a set of three customised mechanical switches. Due to the moving electron source of the MIMOS sensor head (see section 4.10.1) the payload cabin has to touch the sample area with a contact pressure of at least 1 N during the measurement to avoid any movement of the cabin itself, which would falsify the measured spectrum. The three switches, with an activation force of 1 N each, are evenly distributed around the instrument aperture to ensure the contact independently of the contacting surface.

4.9.4 Rotational sensors

The rotational sensors are integrated in the actuator drives on the lever arm. Both drives have a shaft which is attached to the bottom of the flexspline cup, which is the output of the actuator drives. The key tip of the shaft is engaging with an encoder. The implemented device for MRP is a rotational position sensor.

4.9.5 Stereo camera

The stereo camera is situated on the lander. This camera was not part of the MRP project, but of a previous ESA TRP study (Planet Micro Cam – PMC) undertaken by a team lead by the vH&S. A prototype of the camera was developed and was assumed as the baseline for this project.

4.9.6 LEDs

Eight red LEDs are integrated on the payload cabin, one on each corner of the PLC housing. These LEDs can be sequentially powered by the PLC controller, serving for the localisation of the rover with the stereoscopic camera of the lander. For redundancy reasons the LEDs are accommodated on the upper and the lower side of the PLC.

4.10 Integration of the scientific payload – the Geochemistry Instrument Package Facility (GIPF)

In parallel to the development of the rover the ESA TRP project the GIPF project addressed the task of integrating the sensor heads into the payload cabin of the microrover. The Geochemistry Instrument Package Facility (GIPF) consists of two spectrometers and one microimager which were chosen as the rover model payload for the BepiColombo mission. The package represents a complementary set of instruments for the analyses of the Mercury surface. The Mössbauer spectrometer examines the mineralogical information, while the Alpha–Particle X–ray Spectrometer identifies the chemical composition of the sample and the microimager provides information about the morphology and the surface structure.

The miniaturisation of these instruments allows their accommodation inside a mobile platform and thus provides the possibility for in–situ analyses by transporting the scientific payload to the different sample sites.

The two spectrometers have previous flight heritage while the microimager was newly developed during the GIPF project. However the spectrometers had to be slightly modified in order to be accommodated together with the camera inside the limited volume of the rover's payload cabin.

During the GIPF project engineering models of the instruments were built, integrated into a reference payload cab and underwent several functional and environmental tests. The reference PLC is a simple box designed with exactly the same volume as the instrument partition of the rover PLC, in order to provide the same conditions for the instrument accommodation.

(a) MIMOS   (b) APXS   (c) MIROCAM

Figure 4.29: Scientific instruments of the GIPF package designed to be accommodated into the Nanokhod payload cab

4.10.1 MIMOS – Mössbauer Spectrometer

The MIMOS instrument (seen in figure 4.29(a)) is a miniaturised Mössbauer spectrometer used to carry out in–situ examination of ferrous minerals. During the 1990s a space–level MIMOS was developed at the Technical University of Darmstadt. The result was the participation in two Mars mission: as part of the NASA Mars Exploration Rovers (MER) *Spirit* and *Opportunity* as well as the European Beagle mission. The two spectrometers on the MER mission were launched in 2003 and continue to work successfully to the present day [41], [42]. The development of the MIMOS instrument continues at the Institute of Inorganic and Analytical Chemistry from the University of Mainz and at vH&S.

The principle of the MIMOS spectrometer is based on the Mössbauer effect [43]. This is recoilless emission and resonant absorption of γ–rays by certain nuclei in solid materials.

For this effect to occur, the absorbed and emitted γ–rays must have exactly the same energy and thus be resonant. This is only possible if the used nuclei are in a solid lattice with a high enough mass in order not to have any recoil energy. The kinetic energy is reciprocally proportional to the mass m and assuming a constant momentum p, viz:

$$E_{kin} = \frac{p^2}{2m} \tag{4.1}$$

The mass of the solid lattice has to be ~ 1018 times the mass of the nuclei. Even with a high lattice mass it can occur that the energy from emission and absorption causes lattice vibrations (phonons), which prevent the Mössbauer effect from occurring. The probability of recoilless emission and resonant absorption is defined as Debye–Waller factor; this factor is dependent on the crystal temperatures and the γ–ray energy. The Mössbauer effect can be observed with many isotopes, though most instruments use the ^{57}Fe–atoms, limiting the results to iron–bearing samples.

The MIMOS instrument consists of two major parts: One is the electronic boards which contain the power supply, memory and the central processor as well as specific support electronics. These generic electronics can be combined with the generic electronics of the other payload instruments to form a common electronic subsystem [44]. The second part is the actual sensor head, which consists of the detectors, amplifiers and the Mössbauer drive and its associated velocity control electronics.

The MIMOS instrument developed for the NASA mission required a mechanical cover for protection against damage. Furthermore it was equipped with contact sensors to provide the contact to the sample surface. For the GIPF instrument version, the contact sensor is provided by the rover and MIMOS is enclosed inside the payload cab, thus the additional mechanical cover is not needed. Still for the shielding of the electronics a thin metal sheet encloses the instrument. The volume of the MIMOS sensor head is $41 \times 41 \times 81$ mm^3 with an mass of 350 g.

4.10.2 APXS – Alpha–Particle X–ray Spectrometer

The Alpha–Particle X–ray Spectrometer is used for the examination of the chemical composition of the sample. The spectrometer is a development of the Max–Planck–Institut of Chemistry (MPCh), Mainz, Germany. The instrument has already significant flight heritage; a previous development was part of the Sojourner payload on the Pathfinder mission. After these first in–situ measurements on the Mars surface the spectrometer was further improved by the MPCh. The lander of the ROSETTA mission, which was launched in March 2004, carries an enhanced APXS for the chemical analyses of the comet surface. Two APXS spectrometers are successfully in use on the American MER rovers to the present day.

The outside dimensions of the instrument for the GIPF payload were further reduced to $40 \times 40 \times 41$ mm^3. The instrument mass could be reduced to 150 g by removing the housing elements which are provided by the PLC walls.

The APXS uses two different methods for the excitation of X–rays: Alpha–induced X–ray emission and X–ray fluorescence. The ionisation with Alpha–particles works very well on elements with a low atomic mass like Sodium, Magnesium, Aluminium and Silicon. The Alpha–particles remove electrons from lower atomic shells, whereupon electrons fall from a higher shell emitting a quantum of energy. The energy of this quantum is characteristic for each element.

The X–rays sent out from the source are preferably ionising elements with a higher atomic mass such as iron. However, the principle stays the same: The X–rays are removing electrons from the lower shells and the up moving electrons emit characteristic energy quanta.

4.10.3 MIROCAM – Microrover Camera

The MIROCAM as part of the GIPF package is a miniaturised imager. It consists of a CMOS–APS imaging sensor with miniaturised electronics and opto–mechanics. The camera has a lens system with an electro–mechanical drive for focusing and a LED–based illumination unit to provide pictures on the night side of Mercury. The instrument can be operated in a close–up mode and in an infinity mode. The close–up mode provides high resolution and multicolour images of the investigated rock samples, while the pictures with infinity–mode supports the navigation of the rover. The camera is a new development of the DLR Institute for Planetary Exploration, Berlin, Germany. Its dimensions are $20 \times 40 \times 100$ mm^3 with a mass of 150 g. Although the camera is used for the navigation system of the rover the main objectives of the camera are scientific research:

- Gather information of the surface morphology, porosity and variation.
- Provide information on structure and mineralogy.
- Analyse the texture and surface properties.
- Analyse the change of material distribution.
- Support the composition analyses by multispectral imaging.

- Support the interpretation of results from the other instruments (APXS and Mössbauer).

4.10.4 LMS – Laser Mass Spectrometer

An alternative instrument for the APXS Spectrometer which was also suggested for the BepiColombo mission is the Laser Mass Spectrometer (LMS), [45].

The LMS is a miniature laser ablation time–of–flight mass spectrometer suitable for the use of in–situ analyses of the elemental and isotopic composition of planetary bodies under vacuum conditions.

The instrument is designed to determine major, minor and trace element abundances in minerals on a spatial scale of 10 microns. The dynamic range and mass resolution is sufficient to also make useful isotopic measurements. Time–of–flight mass spectrometers are simple, robust devices; the most recent prototype instrument (figure 4.30) has a design mass of 0.25 kg, which is small enough to be integrated in the PLC of the rover.

Figure 4.30: The second LMS prototype built by the University of Bern

The instrument principle is based on the following effect: If an incident laser pulse upon a surface has a sufficiently high irradiance ($\sim 1\,\mathrm{GW/cm^2}$), the local material absorbs energy much faster than it emits it. This evokes a near instantaneous formation of hot dense plasma which expands away from the surface. The plasma conditions allow the ionisation of most elements. The ionised elements in the plasma are then analysed by a mass analyser which is coupled to the laser system. Using the effect of ablation by the laser, the measurements also provide a depth profile of the sample surface.

5 Thermal system Nanokhod

5.1 Thermodynamic principles

Heat transmission takes place in three different ways: transfer by conduction, by convection and radiative heat transfer. These can be separated into two groups: Conduction and convection, which are bound to a medium, while radiation is independent from a transfer substance [46].

For heat conduction the energy is transported by the interaction of bordering molecules and atoms which have an uneven temperature distribution. It is also called intermolecular heat transport.

For convection a macroscopic movement of the particles is necessary for the transport of energy. This is only possible within liquid or gaseous fluids. The convective heat transfer is thus always related to the flowing movement of the fluid. For fluid movements driven by external forces the heat transfer can be forced, which then is called *forced convection*. If the flow of the fluid is based only on natural forces like buoyancy, the heat convection is *free* or *natural*.

Thermal radiation is the electromagnetic radiation which is thermally stimulated and thus depends only on the temperature and the properties of the body that is emitting the radiation. All bodies with a temperature above $0\,K$ emit thermal radiation.

Each body tends to an equilibrium between the energy flux it absorbs from the environment and the energy flux it emits to the environment. These energy fluxes to and from the body can be a combination of all three heat transfer types: conduction, convection and thermal radiation.

For the thermal analyses of the Nanokhod, convection does not need to be considered as the Mercury surface has only a negligible atmosphere with a surface pressure of $\sim 10^{-15}$ bar and the rover system does not contain any fluids.

5.1.1 Heat transfer by conduction

For conductive heat transfer the energy is transported by the interaction between molecules and atoms and thus depends on the contact situation in between those and on the temperature gradient.

According to the Fourier law the heat flux density $\dot{q}$ inside a homogeneous and isotropic body is proportional to the temperature gradient ∇T $\left[\frac{K}{m}\right]$:

$$\dot{q} = -\lambda \nabla T \tag{5.1}$$

The proportionality factor is the thermal conductance λ $\left[\frac{W}{mK}\right]$, which is a material property.

Assuming a wall with large extents in the $y-$ and $z-$direction, the Fourier law can be modified to:

$$\dot{q} = \frac{\lambda}{d}(T_1 - T_2) \tag{5.2}$$

for the heat flux density $\dot{q}$, through a wall of a thickness d. T_1 and T_2 are the temperatures on the two surfaces of the wall.

When relating the heat flux to the wall surface area A, it is given by:

$$\dot{Q} = A\dot{q} = \frac{A\lambda}{d}(T_1 - T_2) \tag{5.3}$$

Summarising the properties of the material and the specific body the thermal conductor h_c can be derived as:

$$h_c = \lambda \frac{A}{d} \tag{5.4}$$

Analogue to the Ohm's law for electronic values,

$$I = \frac{U}{R} \tag{5.5}$$

with the electric current I, the voltage U and the electrical resistance R, the thermal resistance R_c, can be defined as the reciprocal term of the thermal conductor:

$$R_c = \frac{1}{h_c} \tag{5.6}$$

and thus follows for the heat flux:

$$\dot{Q} = \frac{(T_1 - T_2)}{R_c} \tag{5.7}$$

With this analogy, the heat flux can be given also for walls consisting of several layers, which differ in their material properties and/or their geometry.

5.1.1.1 Heat flux perpendicular to the wall layers

A complete wall consisting of n layers with the thicknesses $d_1 \ldots d_n$ which have the heat conductance $\lambda_1 \ldots \lambda_n$ is assumed. The temperature distribution of the wall is given with T_1 on one side of the wall and T_2 on the other side with $T_1 > T_2$. For the overall heat flux through the wall the heat fluxes through the individual layers are added up, giving the following equation:

$$\dot{Q} = \lambda_S \frac{A}{d}(T_1 - T_2) \tag{5.8}$$

λ_S is the heat conductance for the complete wall:

$$\lambda_S = d \sum_{i=1}^{n} \frac{\lambda_i}{d_i} \tag{5.9}$$

For the heat flux perpendicular to the layers, the thermal resistances are connected in series.

$$R_{c_P} = R_{c_1} + R_{c_2} + \ldots + R_{c_n} = \frac{d_1}{\lambda_1 A} + \frac{d_2}{\lambda_2 A} + \ldots + \frac{d_n}{\lambda_n A} \tag{5.10}$$

5.1.1.2 Heat flux parallel to the layering

For the heat flux parallel to the wall layers, the wall consists again of n layers with a thickness of d_i with $i = 1...n$. The layers have cross–sections perpendicular to the heat flux of $A_i = b \times d_i$ and the heat conductance λ_i. The width of the wall setup is b and the length in direction of the heat flux is l. With the temperatures T_1 and T_2 ($T_1 > T_2$) on the two opposite sides of the wall, the heat flux through the complete wall is again given by the sum of the individual layer heat fluxes:

$$\dot{Q} = \sum_{i=1}^{n}(\lambda_i d_i)(T_1 - T_2)\frac{b}{l} \tag{5.11}$$

The heat conductance λ_P for the complete wall is:

$$\lambda_P = \frac{\sum_{i=1}^{n}(\lambda_i d_i)}{d} \tag{5.12}$$

with $d = \sum_{i=1}^{n} d_i$.

For the calculation of the heat flux parallel to the wall layers the overall thermal resistance corresponds to a parallel connection of the individual resistances:

$$\frac{1}{R_{cS}} = \frac{1}{R_{c_1}} + \frac{1}{R_{c_2}} + ... + \frac{1}{R_{c_n}} = \frac{\lambda_1 A_1}{l} + \frac{\lambda_2 A_2}{l} + ... + \frac{\lambda_n A_n}{l} \tag{5.13}$$

5.1.1.3 Contact heat transfer

For the heat transfer in between two components an additional factor h_t, the interfacial conductance, has to be considered. The factor describes the contact conditions between two bodies. This factor depends on a number of different aspects: material properties, the surface roughness, the ambient conditions and the contact force of the two components.

h_t has to be included into the heat transfer equation (see also [47]). The coefficient has the same unit – $\left[\frac{W}{m^2 K}\right]$ – as the heat transfer coefficients of materials. The contact resistance R_{ci} $\left[\frac{K}{W}\right]$ through an interface area A thus results in:

$$R_{ci} = \frac{1}{h_t A} \tag{5.14}$$

The complete thermal resistance of two material layers including the contact resistance is thus:

$$R_c = \frac{d_1}{\lambda_1 A} + \frac{1}{h_t A} + \frac{d_2}{\lambda_2 A} \tag{5.15}$$

The high number of dependencies contained within the heat transfer coefficient h_t makes it difficult to find a specific value in literature for every application. Most often a range for this factor is given which has to be rendered more precisely by test for the specific conditions and applications.

5.1.2 Radiative heat exchange

Especially in space applications thermal radiation has a strong influence on the thermal design. Within the vacuum conditions of space, there is no convection and the conductive heat transfer is limited to the system structure, so radiation is the only way of exchanging heat fluxes through the interspaces. For the example of the Nanokhod, the system exchanges heat by conduction only within the rover and with the Mercury surface, while the rest of the heat exchange is only radiative.

The radiative heat exchange depends on the temperature of the body, its physical emission properties and the geometrical configuration of the emitting and the absorbing body. The emission ratio ε describes the ratio of the emitted radiation compared to radiation which would be emitted by an idealised black body. It is driven by the temperature of the emitting body, the wavelength of the emitted radiation, the angle of incidence and the property of the emitting surface. A black body can be assumed to have an absorption as well as an emission ratio of 1.

$$\varepsilon = \alpha = 1 \tag{5.16}$$

For a non–idealised emitter (i. e. practical), both the absorbing and radiation ratio is below 1, with a proportion of the radiation being reflected from the surface and another proportion transmitted through the body. Nevertheless for all emitters – black and real ones – the following equation applies:

$$\alpha + \rho + \tau = 1 \tag{5.17}$$

where ρ is the reflection ratio (the share of the incident light that is reflected), α the absorption ratio (the share of the incident light that is absorbed) and τ the transmission ratio (the share of the incident light which is transmitted through the body).

The reflection may be specular – the incident ray of light is reflected as one sharply contoured ray – or diffuse – the incident ray is reflected in all directions. The reflection ratio contains both types of reflection. The type of reflection depends on the surface properties; the rougher the surface is, the more diffuse the reflection will be.

The radiation flux follows the Stefan–Boltzmann–Law:

$$M = \varepsilon \cdot \sigma \cdot T^4 \tag{5.18}$$

with the surface specific radiation M, the emissivity of the material ε, the temperature T of the emitter and the Boltzmann constant $\sigma = 5.67 \times 10^{-8}$ W/m^{2}K^4.

The radiation of a body covers the whole wavelength range of the thermal radiation. The radiative intensity though varies over the range of the wavelengths. The wavelength with the maximum intensity λ_{max} depends from the temperature T of the emitting body and can be calculated by Wien's displacement law.

$$\lambda_{max} = 2897.7 \frac{\mu mK}{T} \tag{5.19}$$

The distribution of the radiation intensity over the ranges of the wavelength follows Planck's law (see [47]).

For the radiative heat exchange between two surfaces, the orientation of these surfaces towards each other has to be considered. Depending on the geometric composition only a certain fraction of the emitted radiation can reach and be absorbed by the second surface. This correlation is defined by the radiative view factors which are defined as follows:

The amount of heat flux Φ_i which is emitted by the body surface A_i and which is directly incident to the body surface A_j is given by Φ_{ij}. The dimensionless fraction of these heat fluxes is called view factor:

$$F_{ij} = \Phi_{ij}/\Phi_i \tag{5.20}$$

Assuming the Emitter i is surrounded by a receiver j – i.e. the receiver forms a cavity – the sum of all view factors from i to j is 1,

$$\sum_k = F_{ij_k} = 1 \tag{5.21}$$

The following steps show the determination of the view factors for two surfaces A_i and A_j illustrated by figure 5.1.

The first step is the calculation of the radiant flux $d^2\Phi_{12}$ emitted from the surface element surface dA_1 into the direction of the surface element surface dA_2 [48]:

$$d^2\Phi_{12} = L_1 \cos \beta_1 A_1 d\omega_2 \tag{5.22}$$

with the radiance L_1, the solid angle $d\omega_2$

$$d\omega_2 = \frac{\cos \beta_2 dA_2}{r^2} \tag{5.23}$$

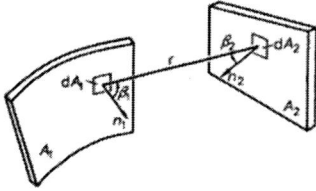

Figure 5.1: Determination of the view factors in between two surfaces

and the geometry shown in figure 5.1 results to:

$$d^2\Phi_{12} = L_1 \frac{\cos\beta_1 \cos\beta_2}{r^2} dA_1 dA_2 \qquad (5.24)$$

Assuming that the radiance L_1 is constant for the whole surface A_1, the equation can be integrated over both areas:

$$\Phi_{12} = L_1 \int_{A_1} \int_{A_2} \frac{\cos\beta_1 \cos\beta_2}{r^2} dA_1 dA_2 \qquad (5.25)$$

Relating this radiation flux to the radiation flux Φ_1:

$$\Phi_1 = \pi L_1 A_1 \qquad (5.26)$$

which is emitted by A_1 into the half space, leads to the view factor that is given by:

$$F_{12} = \frac{\Phi_{12}}{\Phi_1} = \frac{1}{\pi A_1} \int_{A_1} \int_{A_2} \frac{\cos\beta_1 \cos\beta_2}{r^2} dA_1 dA_2 \qquad (5.27)$$

For reasons of symmetry the reciprocity relationship for the radiation flux from A_2 to A_1 is given by:

$$F_{21} = \frac{\Phi_{21}}{\Phi_2} = \frac{1}{\pi A_2} \int_{A_2} \int_{A_1} \frac{\cos\beta_2 \cos\beta_1}{r^2} dA_2 dA_1 \qquad (5.28)$$

This reciprocity related to the view factors gives the following equation:

$$A_1 \cdot F_{12} = A_2 \cdot F_{21} \qquad (5.29)$$

The integration over all surfaces to derive the view factors is a computationally intensive operation especially for geometrically complex systems like the Nanokhod. Special software has been developed to gain the values more easily by using different methods. The ESARAD software used for the thermal analyses of the Nanokhod determines the view factors with the Monte–Carlo ray tracing method or the Matrix method (see section C.2).

For the determination of the heat flow between two surfaces A_1 and A_2 it is thus necessary to know the temperatures T_1 and T_2, the thermo–optical properties which are given with the emission rate of a non–black body ε_1 and ε_2 as well as the view factor F_{12} of the geometric configuration. The emission rates and the view factor are combined to give the transfer factor F_{12}

$$Q_{12} = F_{12}\sigma(T_1^4 - T_2^4) \tag{5.30}$$

or in analogy to conductive heat transfer the equation including the thermal radiation resistance is:

$$Q_{12} = \frac{\sigma(T_1^4 - T_2^4)}{R_{rad}} \tag{5.31}$$

with the thermal radiation resistance R_{rad}

$$R_{rad} = \frac{1 - \varepsilon_1}{\varepsilon_1 A_1} + \frac{1}{F_{12}A_1} + \frac{1 - \varepsilon_2}{\varepsilon_2 A_2} \tag{5.32}$$

5.2 Impact of extreme environments on space application

The BepiColombo mission to the cold side of Mercury encounters an extreme environment. The mission will be exposed to very high thermal radiation due to the high solar flux so close to the sun. With no atmosphere and the very slow rotation period of the planet, the thermal environment varies between two thermal extremes. After some initial study to realise a mission which would encounter both the day and the night side of Mercury, this was decided to be too challenging while adding only a marginal scientific advantage. Even with a mission visiting only the night side of the planet, significant challenges still remain for parts and materials.

Without atmosphere neither convectional cooling nor conduction via a gas atmosphere can occur. This results into very low temperatures on the planets surface but at the same time there is the risk of hot areas on heat dissipating components. Especially the electrical health–checks of the systems during the transfer to Mercury with environmental temperatures up to +70°C are heat critical. In combination with the design approach to minimise the heat losses of the system, which is necessary for the on surface mission phase, the electronic checks can lead to high local temperatures.

Potential problems caused by the space and Mercury environment in general shall be considered in the following sections.

5.2.1 Constraints for parts and processes

Although the mission to the Mercury surface was finally planned to go to the cold side of the planet with −180°C, still some effects of high temperature have to be considered, as electronic parts act as heat sources and partially have low thermal linkages within the vacuum conditions to remove the heat fast enough.

Trends in material characteristics given for the standard temperature range often continue for both higher and lower temperatures. However for some critical temperatures, there may be sudden changes in behaviour. It is thus risky to extrapolate the characteristics curves without monitoring for extreme temperatures.

For high temperature conditions on the electronic boards the following effects have to be considered (see also [49] and [50]):

- Mechanical parts and connections are stressed by thermal expansion.
- The melting point of soldering.
- The glass transition point of dielectrics.
- The curie point of magnetism.
- Electromigration.
- Change of metallurgical properties:
    - Material weakening of electrical and mechanical contacts.
    - Formation of intermetallics particularly with gold.
- The change of electrical resistance.
- Change of reference and threshold voltages.
- Development of leakage currents.
- Change of thermal conductance.
- Changes in propagation delays.

For low temperature conditions the main concerns result especially from the following aspects (see also [51] and [52]):

- Mechanical parts and connections are stressed by thermal expansion.
- Materials become brittle.
- Freeze–out effects for movement of ions or chemical processes (e. g. low capacitance, battery freeze–out).
- Change of electrical resistance (especially for thick–film resistors).
- Change of thermal conductance.
- Changes in component propagation delays and internal gains.
- Increased viscosity and solidification.
- Metallurgical effects.

There are no commercial parts available which are qualified for such a temperature range which has to be considered for the current Mercury mission. Parts tested for the standard military range and ratified as space components are usually qualified for temperatures in between $-55°C$ and $+125°C$ operational range.

In order to identify adequate parts for this mission, technically appropriate components have to be screened by test [39], [53] and [54]. These screenings also bear the risk that the selected parts become obsolete when coming to the next phases of the project. For this reasons, the screening should concentrate on standard parts rather than on "exotic" components.

Helpful for this approach is the definition of areas within the design which are less demanding from the temperature range. These areas can be used to place components which are more difficult to predefine and screen in an early design process. This of course means that a detailed thermal analyses of the system has to be available and has to be kept up to date through the whole design phase.

Nevertheless the choice of parts for such a mission is a long iterative process, which alternates between the search and the testing of appropriate parts, and as the last step, the redesign of the system to accommodate the finally available parts. In order to make this effort as small as possible it is sensible to keep the design flexible and allow for alternative solutions wherever it is possible.

As many effects such as variations of electrical parameters and metallurgical effects depend on interfaces between parts, the testing of suitable components is not sufficient; instead bread–boarding of subsystems and systems has to be carried out to detect such effects.

5.2.1.1 Thermal effects on electrical systems

Temperature gradients have an impact on thermoelectric effects for example by inducing considerable voltages in electric circuits with varying conductor materials or by additional electromigrational effects, diffusion or decrease of thermal conductivities.

Property differences of unequal materials can decrease for high temperatures. For example, materials with high electric resistance, the resistance decreases with increasing temperatures; vice versa materials with a low electric resistance will have an increasing resistance in high temperature conditions. This means engineering and designs based on effects caused by the differences between the materials (e. g. Peltier elements) will lose their function due to the disappearance of such effects in extreme temperature environments.

For magnetic principals the Curie point of materials has to be especially considered, while capacitors, PCBs and structural materials have to be revised considering the glass transition point.

For most structural materials like aluminium, steel, copper or PTFE the change of temperature results in a change of the electrical resistance. Silicon develops due to an increase of free carrier excessive leakage current, for high temperatures starting already at temperatures of $+180°C$. As an alternative, SOI devices (silicon on insulator) can be used, which have an additional insulation layer and thus reduce the high leakage effects for high temperature conditions. GaAs (Gallium arsenide) has wider band–gaps and thus the free carriers and the effects of leakage current only develop at higher temperatures. The band–gap of SiC (Silicon carbide) is even higher than for GaAs and it has been demonstrated to work well up to temperatures of $+500°C$. Cold temperatures however can lead to freeze–out effects, slowing the movements of ions or chemical processes.

5.2.1.2 Thermal effects on materials

Different coefficients of thermal expansion (CTE) of materials induce mechanical stress on component interfaces. The change in the tolerances due to different coefficients of thermal expansion has also negative effects on mechanisms. The CTE of materials changes with the temperature however the CTE characteristics vary for different materials. This is why material combinations have to be carefully selected, considering the whole temperature range, in order to minimise the variation in expansions and thus the thermal stress.

An additional effect is the change of the thermal conductivity for materials, influencing the thermal design.

Temperature variations also change the strength, flexibility and brittleness properties of materials. The importance of material flexibility and the drawback in property changes is obvious in the sealing issue discussed in section 4.3. Change of strength and brittleness properties have to be considered in structural analyses and tests. Both have to be carried out with property values for the realistic temperature conditions.

The outgassing of materials also increases for high temperature conditions, which has to be considered for outgassing sensible subsystems like optics. This is especially critical for plastics like PCB materials or package materials of electrical components.

Even more problems than described for semiconductors are caused by the packaging and bonding of components. Almost all plastic components used for electronics can not be used above $+130°$C. Polyimides like Kapton keep their characteristics up to $+220°$C. PTFE is usable up to temperatures of $+200°$C, but it also works nicely for very low temperatures. The use of ceramic substrates is also advantageous for high and low temperatures. Additionally the ceramic substrate matches mechanically very well with ceramic packages of components. With a glass transition point for most materials between $+130°$C and $+180°$C, the problem might not even be the melting, but the change of their electrical characteristics, which is e. g. true for FR4 (standard PCB substrate material) for $+140°$C.

5.2.1.3 Impact on processes

With the change in length and volume over the temperature range mechanical stress will occur on the mounting points of components. This has to be considered for the design of the PCBs and for the choice on how to mount the components. It has to be also noted that the solder performance is not only a function of the solder but also of the material to be soldered and that its strength and flexibility is highly non–linear over temperature.

Indium solder alloys provide ductility, which absorbs to some extends the different rates of expansion and contraction of materials with unlike CTEs without cracking the bond.

Temperature effects on processes can also result from the interaction of different materials with each other. One example is the combination of gold and standard tin solder resulting into a highly brittle gold compound, which is risky especially for electronic connections.

5.3 Thermal design of the Nanokhod rover

Especially in small highly integrated systems the impact of different requirements and of all other subsystems is very high. During the design phase the thermal control system is only one of the factors. When optimising the overall system, trade–offs between different subsystems have to be made and the thermal control system itself will not necessarily be optimal.

The Nanokhod rover has a mainly passive thermal control concept. The only active parts in the system are some heating resistors which heat specific drive electronic components in order to ensure their correct operation. Apart from that, the thermal system of the rover relies on the design and integration, which was made to keep the heat within the system as much as possible, whilst avoiding single hot areas.

A passive thermal control system has the advantage of no additional increase of the system's power consumption, which is beneficial for the considered mission. Another advantage is the positive effect on the mass resource, as only limited additional system mass is necessary opposing to an active system [56]. However, it is less effective and in general slower in responding the development of hot or cold areas.

Due to the small size of the system and the resulting compact design of the rover, it can be assumed that under atmosphere conditions the thermal gradients within the rover are rather low. This would mean that the thermal equilibrium is reached within a short time, but due to the vacuum conditions, the energy transfer via conduction and convection through the atmosphere is missing. The small rover size leads to extreme temperature changes as the system itself only provides a very limited heat capacity. If the conduction via the rover structure is too low, single components may develop too much heat. This procedure is supported by thermally isolating materials combined with heat dissipating components and parts with a very low heat capacity which locally influence the thermal system.

In order to avoid these hot areas and, on the other hand, the high losses of thermal energy, the design of the rover bases on the following two principles: The rover is designed to work in a $-180°C$ –environment without additional heating. At the same time the rover retains as much heat as possible, resulting from dissipation, inside the rover. These guidelines are realised within the design of the rover as follows:

The thermal conduction from the locomotion units of the rover into the regolith surface was minimised by implementing a PTFE sealing membrane in between the locomotion unit body and the track foil (see also section 4.3). With the low thermal conductance of PTFE the energy losses over time can be reduced.

In order to also reduce the losses via thermal radiation from the rover surfaces to the environment all aluminium surfaces of the rover structure are coated with Alodine 1200. Alodine has a very low emissivity of approximately $\varepsilon =0.12$. The surface treatment with Alodine provides corrosion prevention and keeps the low emissivity of polished aluminium. The treatment, also called aluminium passivation, contains hexavalent chromium and has a very low electrical resistance. The coating leaves the surface with a coloration ranging from a light Iridescent gold to tan depending upon alloy and immersion time.

In order to avoid hot spots on the PCBs a good thermal linkage towards the main rover structure is required. For this reason, all boards populated with the heat dissipating electronic components are mounted on aluminium standoffs. The standoffs are optimised to have a big enough contact surface for a good thermal conduction without taking to much area on the boards needed for the accommodation of the electronics. The length of the standoffs is minimised for an optimal heat removal from the boards, but still provide the necessary height above and below the PCBs for the accommodation of the electronic components.

For the control of the thermal status the temperatures of critical electronic components are monitored in order to immediately note an overheating of these areas and to be able to switch of the system and thus avoid any damage. In addition to that on–board control of the temperatures on critical components during the later mission, the thermal design also undergoes a thermal testing campaign in which these aspects are analysed and validated. During the thermal–vacuum tests the modes with a high thermal dissipation are conducted in order to test the capabilities of the thermal design.

5.4 Thermal requirements

With the thermally demanding environment of both the transfer to the planet and the Mercury surface the mission requirements impose a demanding task for the design of the rover thermal system.

The requirements having the main impact on the thermal design are summarised here below. For the transfer of the rover from earth to the Mercury orbit the rover system has to undergo with temperatures of $-150°C$ to $+70°C$. These temperatures are for most of the components non–operational temperatures, but all electronics will be tested during the flight and thus also have to work within this range. All components of the system have to be designed to operate starting from the expected Mercury surface temperature of $-180°$C upwards and under vacuum conditions. The solar irradiance on the night side of the Mercury surface is expected to be 0 $\frac{W}{m^2}$, which means that also the Albedo radiance is zero.

The Mercury surface is expected to consist of very fine regolith similar to the moon, which has a very low thermal conductance. The energy supply of the rover by the lander is limited to 265 Wh, which also limits the power for the preheating of electronic components. The internal heat dissipation depends on the different operational modes of the rover during the mission (see section 4.8).

The most significant aspects of thermal heat transfer for the Nanokhod system are:

- Conductive heat exchange between the track foils of the locomotion units towards the regolith.
- Radiative heat exchange between the rover and the regolith surface.
- Radiative heat exchange of the rover with deep space.
- Internal heat dissipation from the rover electrical components to the surrounding structure.

6 Thermal modelling of the Nanokhod rover

For a complex system like the Nanokhod rover, a simple thermal estimation will give some initial numbers and an idea in which magnitude the temperatures will range for the given environmental conditions. In order to get representative limit values for the electronic parts or the scientific instruments, it is essential to model the rover system and its environment to more detail.

During the development phase of the rover the thermal mathematical model was developed in parallel and became more refined with the progressing of the project. The first model was a simple spread sheet calculation giving approximate magnitudes of temperatures. The second model was built with the ESATAN–/ ESARAD–Software supported by a compiler tool *Design Compiler 43*. Based on the second model there have been several iterations on the thermal network. The iterations considered areas of special interest by refining the network on this part, either because they showed unexpected temperatures or because they were thermally sensitive parts. The thermal model was also iterated to predict rover temperatures for the different environmental conditions.

The rover breadboard was designed with the intention of performing thermal–vacuum testing. The thermal rover model was thus reconfigured to the environment of the thermal chamber. The test results gained during the thermal–vacuum testing was then used to evaluate the model. Deviations in between the temperature results of the test and the model caused by material property assumptions and simplifications in the model were then ameliorated or even completely resolved.

6.1 Simplified thermal model using a six–node evaluation

The simple thermal model consisted of 6 rover nodes for which temperatures were calculated plus one planet surface node with a defined temperature. The nodes relate to the following rover parts: Payload cabin, Tether unit and one inside and one outside half for each locomotion unit.

The thermal paths and accordingly the thermal resistances were assessed between all nodes using the known material properties and a simplified design geometry. As for all following models the heat generated within in the rover was calculated from the rover power budget by removing the mechanical power output and assigning the dissipation to the specific nodes. The total power dissipation for the rover for different operational modes is given in table 6.1.

Activity	Dissipated power [W]
Localisation	2.29
Movement	3.48
PLC positioning	3.48
Science APXS	1.86
Science MIROCAM	2.38
Science MIMOS	0.88
Idle	0

Table 6.1: Total power dissipation for the different operational modes of the rover

The model gives some initial transient times to estimate how fast the heat is dissipated, if the rover is disposed to the environment.

6.2 Thermal network analyses model

In the first network solution for the thermal model, each main structural part was defined as a single node of the thermal system with homogeneous material properties. The resolution of the thermal network thus gave a single temperature for each component. For some of the parts this is a rather rough resolution, while some smaller components are already modelled to a quite high detail. However this approach gives a good initial overview of the temperature distribution in the rover system.

6.2.1 Environmental conditions and requirements

With the preparation of the thermal–vacuum tests in parallel, the thermal model was prepared for the environment of the thermal chamber. The thermal boundary conditions were defined for the setup of the rover inside the chamber, while the rover is situated on a cooling plate and covered with a shroud. These two boundary components are considered to be at the same fixed temperature and represent the radiation environment for the model, as well as the conductive interface for the LUs.

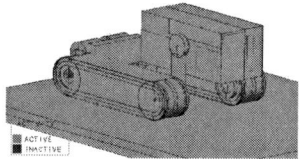

Figure 6.1: Geometry of the rover model

6.2.2 Assumptions and simplifications

The geometrical model designed with the thermal analyses software displays the parts as primitives like cubes and cylinders, see figure 6.1. These primitives can be integrated with each other to realise more complex geometries. Nevertheless a full implementation of the rover geometry in this model will be exceedingly complex, whilst not having a major advantage concerning the determination of the thermal behaviour of each part. Some of the parts geometries are thus simplified concerning its shape, while retaining an accurate volume and correct contact interfaces to its surroundings.

6.2.3 Utilised software for thermal modelling

The thermal model is built with the ESARAD and ESATAN software. For the purpose of a simpler user interface as well as an automated calculation of the heat conduction, the model development uses the *Design compiler 43* developed by IILS mbH in cooperation with the University of Stuttgart and further industry partners. The compiler uses *Mathematika* as mathematic solver. The *Design compiler 43* is used for the definition of the single component's geometry, the connections in between them and it includes the properties that are necessary for the heat transfer calculations. The use of the compiler simplifies changes of the rover model or the environment adding flexibility to the model application. The model geometry is then translated into Fortran 77, which is the programming language used by ESARAD and ESATAN.

For radiative analyses of the geometry it is imported by the ESARAD GUI. The ESARAD programme is used to determine the view factors between the surfaces of the single nodes defined in the model geometry. From these factors the programme then calculates the radiative couplings.

The output of ESARAD combined with the conductive couplings calculated in the compiler with the *Mathematika* plug–in is a mathematical model of the rover in the defined environment which is imported by the ESATAN programme. ESATAN is the thermal analyses programme that solves the thermal network of nodes, conductances, radiation and material properties. Depending on the selected solver routine the output provided by ESATAN is a steady state or transient temperature solution using the thermal network technique.

ESARAD is the standard radiative analyses tool of the European Space Agency and was kindly provided together with the ESATAN software by Alstom, Whetstone, UK, for the thermal analyses of the Nanokhod in the framework of this thesis.

A schematics of the thermal model design flow is given in figure 6.2. The thermal model software, ESARAD, ESATAN and the *Design compiler* and their use are described in more detail in the appendix C.

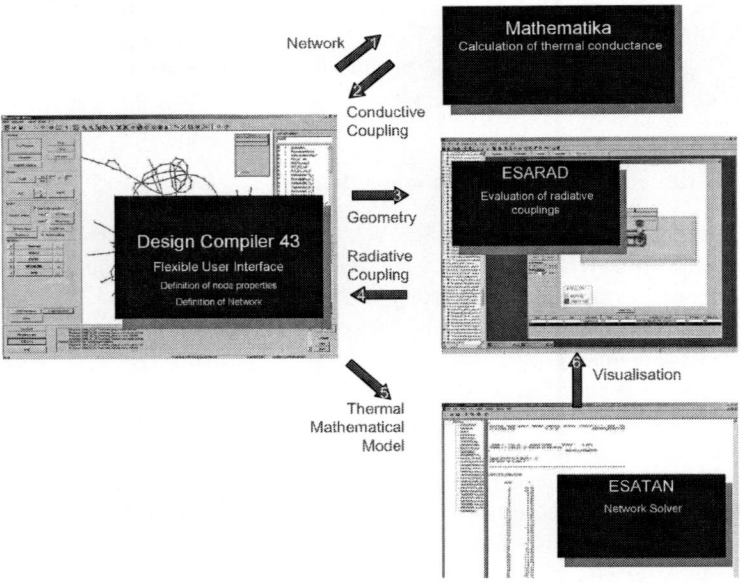

Figure 6.2: Thermal modelling software design flow

6.2.4 The main aspects of the thermal model

For the thermal model of the Nanokhod, all rover components are integrated into a geometric model which was included in the ESARAD programme. The model contains the geometric information and also all material and surface properties, as well as the interface information in between the nodes. Each mechanical rover component consists of at least one node. Large or thermally critical components are divided into several nodes.

Although the rover model is already much more precise than the first simple model, it was still necessary to use simplifications in the model: The components are built from a set of primitives which thus only approximate the actual component geometry.

The geometrical modelling principle mainly focuses on representing as accurately as possible the outside surfaces of the components, their properties and their orientation towards each other, which are important for the radiative heat exchange. ESARAD uses this geometrical model only to determine the view factors within the rover and this determines the radiative heat exchange. The conductive heat transfer is calculated separately based on additionally supplied information relating to wall thicknesses, contact surfaces, component's mass and further material properties. Realistic values for thermal conduction depend on the accuracy of this information.

6.2.5 Modelling results

The first performance of the thermal model calculation was performed before the first thermal–vacuum test and had two objectives:

The first objective of this model was to show that the planned functional rover tests can be safely conducted also for room temperature conditions. With the chamber boundary conditions and the heat dissipation of the active electronics implemented, the results of the thermal model showed that the temperatures would not exceed critical levels which may damage any component.

The second objective was to use the results of this model for comparison with the results of the first thermal–vacuum test. The comparison of the model and the test data allows the validation of the thermal model and can be used as an input for the next model iteration.

6.3 Testing of the thermal conductance of the rover PCBs

With the varying PCB types implemented in the rover system, the use of one thermal conductance value was a significant simplification to represent the thermal behaviour of all electronic boards. The conductance and the specific heat capacity of the PCBs used for the first thermal model, were derived from the material value of FR4 and copper and ignored the number of layers and the actual amount of copper included in the PCB design. Due to the difficulty to calculate the correct value with all the circuit paths on the boards, the better way to gain more realistic data is to measure the thermal conductance and specific capacity of the rover boards.

6.3.1 Test setup

The tested PCBs are isolated from the environment with foam material to reduce the heat losses via radiation and conduction to the surrounding atmosphere. The wrapped boards are placed on top of an aluminium plate (see figure 6.3), which is cooled by 4 aluminium cold fingers reaching down into the silicone oil (Kryo 51) of the heat exchanger. The heat exchanger enables the temperature of the heating plate to remain at a fixed level during the test.

Figure 6.3: PLC controller PCB installed on the cooling plate

The temperatures of the PCBs and the setup were recorded with several PT100 sensors. One thermistor was placed on the cooling plate, while the other sensors recorded the temperatures of the boards in different locations. All the electronic boards were equipped for the tests with a heater resistance R of $1\,\Omega$ in order to provide a controlled heat source. With a current I of $1\,A$ applied on the resistor, the generated heat flux resulted to $1\,W$:

$$P(t) = \dot{Q}(t) = U(t)I(t) = I(t)^2 R \tag{6.1}$$

The test setup was additionally shielded from the environmental heat radiation with a shroud. The shroud surface was laminated with a MLI foil providing good reflectivity of heat radiation.

The test procedure for the PCB test was as follows:

1. Set the silicone oil of the heat exchanger to $-10\,^{\circ}$C and cool the plate to steady state temperature.
2. Install PCB and cool to steady state temperature.
3. Supply current to the resistor for its use as a controlled heat source and leave the system to dwell until the new steady state temperature is achieved.

6.3.2 Determination of the thermal conductance

The heat flux $\dot{Q}$ through a body with the cross–section A and the temperature gradient $\frac{dT}{dx}$ along the PCB dimension x can be determined with the following equation:

$$\frac{dQ}{dt} = -\lambda A \frac{dT}{dx} \tag{6.2}$$

with the heat conductance $\lambda \left[\frac{W}{mK}\right]$.

This equation assumes that the radiation and the conduction to the surrounding atmosphere are negligible and that the PCB is fully isolated from the environment due to the foam material enclosing the boards. Based on the known heat flux $\dot{Q}$, the heat conductance of the PCBs can thus be derived as:

$$\lambda = \dot{Q} dA \frac{1}{\Delta T} \tag{6.3}$$

with the given distance between the thermistors d, the cross–section area A of the PCB, and the temperature difference ΔT resulting from the measured temperature on the resistor T_1 and the one near the standoff towards the cooling plate T_2. The measured results for the heat conductance values are given in table 6.2.

6.3.3 Determination of the heat capacity

The specific heat capacity c_p of the PCBs can be derived from:

$$\dot{Q} = c_p m \frac{dT}{dt} \tag{6.4}$$

with the PCB mass m, the mean temperature gradient $\frac{dT}{dt}$ and the heat flux $\dot{Q}$, being a share of the cooling power provided by the heat exchanger. The remaining share of the cooling power are the heat losses of the setup.

$$P_{\text{eff}} = \dot{Q} + P_{\text{losses}} \tag{6.5}$$

The effective cooling power of the heat exchanger is P_{eff} =0.85 kW at a temperature level of $-10\,°C$. To determine the ratio of the heat flux $\dot{Q}$ to the complete cooling power P_{eff} a reference body with known properties is used. This aluminium body has a mass of 280 g, the specific heat capacity c_p =896$\frac{J}{kgK}$, the heat conductance λ =205$\frac{W}{mK}$, the thickness d =15 mm and cross−section of A =7100.8 mm^2.

With the known data of the aluminium body and the measured temperature difference dT =11.9 K over a time of dt =600 s the cooling flux can be calculated to $\dot{Q}$ =5.01 W. With the determined heat flux, the heat capacity of the different PCBs can be calculated with the measured and known values given in table 6.2.

PCB	Temp. re-sistor T_1 [°C]	Temp. standoff T_2 [°C]	Mass [kg]	d [m]	A [m^2]	Heat flow $\dot{Q}_{\text{res}}$ [W]	λ [W/mK]	c_p [J/kgK]
PLC contr.	53	21	0.0098	0.028	9E−05	1	9.68	310.16
Motor driver	57	38	0.0029	0.013	6.2E−05	1	11.32	636.23
Motor contr.	51	32	0.0055	0.020	6.2E−05	1	16.45	367.19
TDU	47	28	0.0084	0.029	6.2E−05	1	24.57	242.69

Table 6.2: Material data tests of the rover PCB's

The influence of the copper layers in the boards has an obvious impact on the conductivity and capacity results shown in the last columns of table 6.2. The tether data unit PCB (TDU) is a 4–layer board and contains the highest amount of copper and thus has the highest thermal conductivity and the lowest specific heat capacity. All other boards are 2–layer boards, with less copper and thus a lower thermal conductivity and a higher specific heat capacity.

6.4 Sensitivity analyses for different modelling aspects

The parameter data used for the thermal modelling is mainly based on values from literature and reasonable assumptions. However, not only the geometry of the components but also the material properties of several components have to be approximated within the model. Material properties are often given as a range depending on different conditions. In addition components such as the PCBs, the laser, the drives and the bearings are a composition of several different materials. Thus, a simplification is made by choosing only one material property even if the value is a combination of the composite properties.

The contact definition between components is also difficult to define as it depends on a variety of factors. Literature generally provides a heat transfer coefficient range to cover all cases.

Furthermore the method of analysis, such as the number of rays per second for the ray tracing, influences the results.

In order to know the impact of these factors and their influence on the result, a sensitivity analysis on several aspects of the modelling was performed. The sensitivity analysis was carried out for a steady–state calculation with the rover in the operational mode *PLC–positioning* (see also table 6.1).

6.4.1 Influence of the thermal conductance of Aluminium

The difference in heat conductance of the rover PCBs was investigated in section 6.3 providing realistic values for their properties to a reasonable accuracy. However the other material properties have not been evaluated by test and are based on literature values.

The main material used for the Nanokhod design is aluminium type 3.1325; the aluminium material data has therefore the biggest influence. The heat conductance for aluminium 3.1325 is given with the range of 130 to 170 $\frac{W}{mK}$ [57]. For the general thermal modelling the value of 130 $\frac{W}{mK}$ is used. For the sensitivity analysis of this parameter, it is set to a 170 $\frac{W}{mK}$ and the results are compared to the initially calculated temperatures. The increased conductance causes the heat to be distributed faster in the rover system and thus to radiate faster to the environment from the outside surfaces. The steady state temperatures of the components thus decrease.

NODE	Node numb. #	Temp. T with $\lambda_{alu} = 130 \frac{W}{mK}$	Temp. T with $\lambda_{alu} = 170 \frac{W}{mK}$	Temp. difference ΔT
PLC instr. side wall	6	−86.79	−88.01	1.22
PLC motor	22	−76.19	−78.80	2.61
PLC contr. PCB	23	−26.19	−27.98	1.79
LLU PLC motor	29	−94.49	−96.22	1.73
LLU trackfoil below	32	−157.38	−157.52	0.14
LLU PLC motor driver	45	−19.13	−22.77	3.64

Table 6.3: Temperatures for varying aluminium conductivities λ_{alu}

The comparison of the steady state temperatures for environmental boundary conditions of $-180°$C for some selected rover components are listed in table 6.3. It shows the effect of the aluminium conductance parameter. Components like the payload cabin side wall on the instrument side and other components, which have a long conduction path to the cool surface, only differ by slightly more than 1 K. The steel track foil, which is in direct contact with the cool Mercury surface, has only a difference of a few tenth of a Kelvin. However, for the heat dissipating parts such as both lever motors the variation is almost 2 K and for the motor driver of the LLU lever motor it is even more than 3.5 K. These differences in temperatures should be covered by the risk assessment margins and thus do not represent a major risk for the mission. However these should be monitored during the future design process, especially for areas with critical components.

6.4.2 Influence of contact conductance h_t

The contact transfer coefficient is usually rather difficult to determine. All aluminium surfaces within the rover are machined with a low surface roughness of *RZ16* and the majority of aluminium components is screwed to each other resulting to a good contact pressure. The contact transfer coefficient should be quite high and have only marginal influence on the thermal resistance. Nevertheless the vacuum in the interstices in between two facing components still reduces the conductance. Literature [47] gives the contact transfer coefficient for aluminium surfaces with a medium contact pressure as $2200-12000 \frac{W}{mK}$, while it reduces for a low contact pressure and evacuated interstices to $100-400 \frac{W}{mK}$. With most of the components screwed to each other, providing a good contact pressure, but still with the interstices evacuated, the contact transfer coefficient is assumed to be $1500 \frac{W}{mK}$. For the sensitivity analyses the value is set to $750 \frac{W}{mK}$ which bases on the assumption of a lower contact pressure or rougher surfaces.

NODE	Node numb. #	Temp. T for $\alpha = 1500 \frac{W}{m^2 K}$	Temp. T for $\alpha = 750 \frac{W}{m^2 K}$	Temp. difference ΔT
PLC instr. side wall	6	−86.79	−81.78	−5.01
PLC motor	22	−76.19	−68.5	−7.69
PLC controller PCB	23	−26.19	−16.45	−9.74
LLU PLC motor	29	−94.49	−86.82	−7.67
LLU trackfoil below	32	−157.38	−152.96	−4.42
LLU PLC motor driver	45	−19.13	−4.35	−14.78

Table 6.4: Temperatures for varying contact transfer coefficients α

Table 6.4 shows again the comparison of the temperatures for the two transfer coefficients for some selected nodes. This time the temperature difference is quite notable for the two different conditions. The temperatures vary even for the non heat dissipating components by ~ 5 K. For the heat dissipating components the difference rises to nearly 15 K. These variations demonstrate significant risk, as they cover already large extents of the margins defined by the ECSS [58]. The ECSS states that the uncertainty for components with unverified thermal control is 15 K added to a residual thermal design margin (which was not yet defined within the BepiColombo mission) and 10 K environmental design margin (test condition tolerances not included).

The fact that the thermal environment covers a temperature range of 200 K also requires the careful examination of this parameter, as the thermal shrinkage of the materials will influence the contact pressure of the components. This means it is advisable to analyse this parameter by measurement for this assembly and for the target temperature, especially for the heat dissipating parts.

6.4.3 Influence of the number of rays in MCRT

The number of rays for the ray tracing procedure to determine the view factors can have an impact on the result. Too few rays lead to inexact solutions while too many rays consume too much calculation time. For the determination of the view factors with ESARAD all nodes have the same *normal* criticality, which means the same number of rays is traced from each node, independent from the size of the surface area.

In order to ensure that the selected number of rays of $n = 10000$ is sufficient for all surfaces inside the rover the view factors of the same model are determined with ray numbers of $n = 100000$ and $n = 1000$ for comparison.

NODE	Node numb. #	Temp. for 10000 rays	Temp. for 100000 rays	Temp. for 1000 rays	Temp. diff. $\Delta T_{10000-100000}$	Temp. diff. $\Delta T_{10000-1000}$
PLC instr. side wall	6	−86.79	−86.82	−86.82	0.03	0.03
PLC motor	22	−76.19	−76.23	−76.23	0.04	0.04
PLC controller	23	−26.19	−26.16	−26.26	−0.03	0.07
LLU PLC motor	29	−94.49	−94.51	−94.51	0.02	0.02
LLU trackfoil below	32	−157.38	−157.38	−157.37	0	−0.01
LLU PLC mot. driver	45	−19.13	−19.13	−19.33	0	0.2

Table 6.5: Temperatures for varying MCRT ray numbers

Table 6.5 shows the temperatures for the different cases. The temperature differences are marginal and thus negligible. The chosen ray number of 10000 is reasonable and could even be reduced to 1000 for the small surfaces of the rover components.

7 Thermal testing of the Nanokhod

7.1 Objectives of thermal–vacuum tests

The main objective of the thermal–vacuum test (TV–test) is to prove the compatibility of the MRP design with the thermal–vacuum environment. Although the facility used for the TV–test is not able to reach the required temperature of $-180°C$, it allows to test the rover system on reduced temperature levels as a first qualification step. The second objective is to provide thermal information for comparison with the results of the thermal mathematical model realised with ESARAD/–TAN [59] to validate the model. The thermal–vacuum test involves the operation of the PLC mechanisms with the rover static on its tracks. There are two separate tests in order to gain as much data as possible from the rover. The two tests allow a larger amount of test data as well as the possibility to have a more detailed look at temperature distributions of specific components during the second test following the evaluation of the first test data set.

The test procedure for both thermal–vacuum tests is as follows:

1. Install rover on the cold plate of the chamber.
2. Run thermal–vacuum functional test to check rover & test setup.
3. Evacuate the test chamber.
4. Conduct several thermal–vacuum functional tests with operation on PLC actuators at different temperature levels.
5. Set chamber to atmospheric pressure.
6. Run functional test to check rover.

7.2 Thermal–vacuum test facility

The thermal–vacuum tests are performed using the thermal–vacuum chamber at the Institute of Space Systems (IRS) of the Universität Stuttgart.

7.2.1 The thermal test chamber of the IRS

The thermal–vacuum chamber (figure 7.1) used for the test has an attainable vacuum of $p = 2\text{–}3\times10^{-4}$ mbar using two combined evacuation pumps. The primary pump is a rotary vane pump going down to pressure levels of 1×10^1 mbar. The secondary pump is a turbo molecular pump achieving the minimum pressure as stated. The chamber facility includes a circulation thermostat (LAUDA Proline RP 855) attached to a tube circuit welded to a cooling plate. The thermostat is able to provide temperatures between $-55°$C and $+90°$C. The length of the chamber is $l = 2$ m with a diameter of $d = 1$ m.

Figure 7.1: The thermal–vacuum test chamber of the IRS

7.2.2 Test setup interfaces

During the test the rover is placed on top of the cooling plate, which is made of copper and has a surface area of 400×600 mm^2. The plate has a pattern of thread holes to allow for a good thermal contact between the test equipment and the cooling plate. However for the rover test the thread holes are not used, as the intention of the test is to reflect the conditions of the Mercury surface as realistically as possible, with the loose contact provided by gravitational forces.

The rover is shielded from radiation of the ambient environment by a shroud covered with MLI foil. The shroud is mounted onto the cooling plate and was thus at the same temperature level as the plate. The cooling plate temperature is also considered as the external radiative heat exchange temperature.

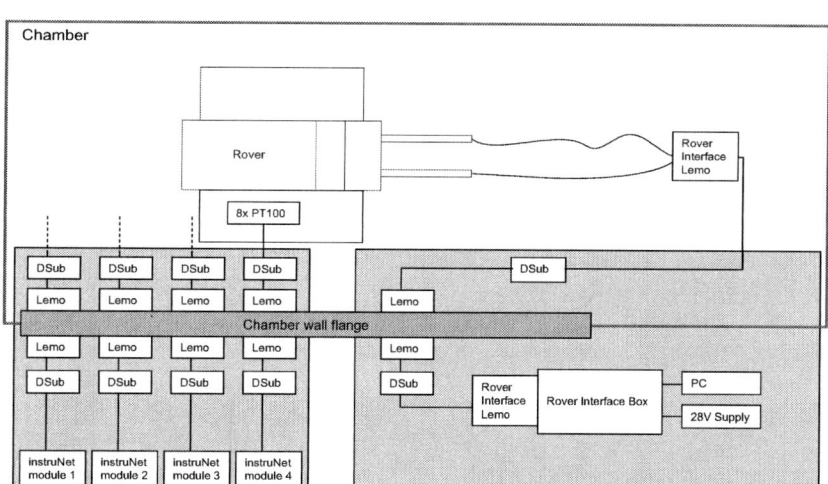

Figure 7.2: Schematic of the electrical feed–through harness for the thermal–vacuum–tests

The electrical feed–through harness consists of two pairs of DSub25 connectors which pass through the chamber wall via two Lemo connectors installed in the chamber flange. The harness feed–through schematic of the test setup are given in figure 7.2. One connector pair (X20/X21) passes via a 4–pin and a 12–pin connection designated for power supplies. The second pair (X30/X31) is intended for the data connection and uses one 24–pin connector for the feed through. The tether cable of the rover requires only a 4–pin connector which is provided by the 4–pin connector of the first feed through DSub25 connector pair. The test setup requires two adapters from the 4–pin to DSub25 connectors: From the rover tether to the chamber feed through and on the outside of the chamber to connect the rover to the interface box.

The wire length for the data and power line as well as for the facility thermistor cables (see section 7.3.5) has to be approximately 1 m. This allows enough slack to move the cooling plate setup towards the chamber door for the installation of the equipment.

7.3 Thermal testing activity

As part of the thermal–vacuum testing activity, several issues have to be considered and shall be described in this section.

7.3.1 Handling precautions

In order not to damage the rover during thermal–vacuum test activity the following handling precautions had to be obeyed:

- Use of gloves to keep rover surfaces clean.
- Electrostatic discharge (ESD) Protection – use a grounded ESD strap.
- The MRP rover is not flight hardware and as such does not require special handling precautions. However, care should be taken not to force the PLC into position and not to damage the track seals.
- For any prolonged duration the unit should be kept in a dry atmosphere (e. g. in sealed container with fresh silica gel desiccant), due to the sensitivity of the dry lubricants against humidity.
- Care should be taken not to deploy tether wires unnecessarily. Any deployed tether wire should not be kinked or broken.

7.3.2 Rover thermal–vacuum functional test

During the thermal–vacuum test, the rover performs a functional test within the test chamber, which tests both the mechanisms and the electronics of the rover. During this test all parameters from the rover housekeeping are logged. Additional measurements are taken during the test as described in the test script IRS2.tc for the first test and in IRS–2008–06–10.tc for the second test (as described in MRP–TN–440–TV–test [60]). The time of each step is recorded from the EGSE so that events and data can be correlated. The rover is supplied during the test from a stabilised 28 V source.

The thermal–vacuum test procedure is described below:

1. Power up rover.
2. Power up locomotion subsystems including motor controller and heater.
3. Power off locomotion subsystems.

4. Power up PLC positioning subsystems including motor controller and heater.
5. Test PLC drive system by driving it for 180° (±90°).
6. Power off positioning subsystems.
7. Test laser – Power on, then off.
8. Test LEDs – Power all on, then off.
9. Power off all subsystems.
10. Power off rover.

7.3.3 Installation of the rover

For the installation in the thermal–vacuum test chamber the following steps are undertaken:

- **Prepare rover:** Articulate rover so that PLC is in starting position.
- **Install rover:** Place rover on the chamber plate. Connect the tether wire to the chamber feed through and perform a functional test.
- **Install facility thermistors:** Install the thermistors as described in table 7.1.
- **Run thermal–vacuum functional test:** With the chamber door open, check if harness collides with mechanisms.
- **Sensor Calibration – step 1:** Calibrate two additional thermistors (One PT 1000, one PT 100) in the circulation thermostat bath.
- **Sensor Calibration – step 2:** Place the calibrated sensors inside the chamber. Evacuate the chamber and leave to dwell. Calibrate all other sensors after reaching an equilibrium temperature.

7.3.4 Thermal–vacuum test configuration

Both thermal–vacuum tests use the same setup of the Nanokhod in the thermal–vacuum chamber. The rover is standing in a half deployed configuration with the track units directly on the cooling plate, which means that the lever arms of the PLC are inclined 45° with respect to the PLC and the LUs. This allows a safe forward movement of the PLC through an angle of 90° during the functional test.

The laser points towards the front of the rover to detect the reflection of the laser light for the visual control during the first test. The locomotion units will stay idle during the functional test inside the test chamber. The MRP drive system will be tested by actuating the positioning drives of the PLC.

7.3.4.1 Test configuration for the first test run

For the first test, the rover is covered with a single shroud cooled by the cooling plate to set the radiation environment on the same level as the cooling plate. A gap of 12 cm in height on one side of the shroud allows visibility of the rover activities during the functional tests. The cooling plate is cooled down in several steps to the minimum temperature of $-40°C$. Since heat is dissipated within the rover and it is also designed to retain heat within the chassis, its temperature is higher than the minimum temperature. In all cases the temperatures of the cold plate surface, the chassis and internal rover temperatures are recorded.

7.3.4.2 Test configuration for the second test run

The test setup of the second test is basically the same as for the first one described above. The slight changes in the configuration of the second test are specified as follows:

The rover is covered with two shrouds which have conductive heat exchange with the cooling plate. The second shroud is inside the original shroud from the first test. The shielding with the extra layer allowed the inner shroud to achieve lower temperatures down to the cooling plate level which was not achieved during the first test. For this test the shrouds are closed to provide a completely cold radiation environment. In order to check upon the rover activities the PLC housekeeping rate is increased to continuously read out the encoder values to be aware of the PLC movement during the test.

During the second thermal–vacuum test the lever actuators are activated for a longer time in order to monitor the temperature developments for the maximum expected activity duration of the rover. The longest planned continuous activation of the actuators is driving a 40 cm path segment in approximately 8 min (based on the rover speed of 5 m/h). The locomotion drives can not be moved during the thermal–vacuum test. Instead the 8 min operation is realised by rotating the PLC 4 times back and forth by $90°$.

The drive configuration for the lever is to a large degree identical the same for the locomotion actuators. The setup of the motor and motor controller for the lever drive in the front of the left locomotion unit is also comparable for the thermal linkage, which allows to draw first conclusions for the maximum temperatures that have to be expected from the long term activity.

7.3.5 Facility thermistors

The measurement equipment of the facility is based on an iNET PC measurement system, which provides four measurement modules. The electrical connection to the PT 100 thermal sensors is via four 44–pin DSub connectors which are installed inside the chamber. On the outside of the chamber the sensors are connected to the four instruNet modules. Each module can connect up to eight PT 100 sensors, which results in a maximum of 32 thermistors for the testing. The thermistor setup is shown in the TV–test schematic in figure 7.2.

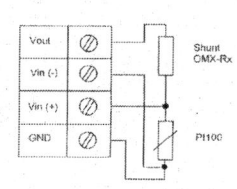

Figure 7.3: PT 100 thermistor configuration

The thermistors are connected to four wires as shown in figure 7.3, connecting them to ground, V+, V− and V_{out}. The 4–wire measurement minimises the resistance error of the thermistor cables and compensates for differences in cable length. Two of the cables, the *force* leads, are connected to the current source, the other two, named *sense* leads, are measuring the voltage. The sense leads are high–impedance leads in order to minimise the voltage drop over the harness in the sensing circuit.

7.3.5.1 Facility thermistors in the first test run

The positioning of the thermistors for the test of the rover is described in table 7.1 and shown in the figures 7.4, 7.5, 7.6, 7.7, 7.8 and 7.9.

Two facility sensors (one PT 1000 and one PT 100) are calibrated before the test in the bath of the circulation thermostat. The calibrated sensors are then mounted

ID	Module Nr.	Location	MRP Module	Comment
J1	M1C1	On PLC wall	PLC	On rover thermal reference point
J2	M1C2	On PLC wall	PLC	Near driver electronics
J3	M1C3	PLC instr. side top lid near LED	PLC	
J4	M1C4	Outer side wall near front yoke	LOU	
J5	M1C5	Outer side wall near rear yoke	LOU	
J6	M1C6	Outer side wall close to lever motor	LLU	
J7	M1C7	On top of the track surface	LLU	Near J12
J8	M1C8	On track surface lower side	LLU	Near J12
J9	M2C7	Inner side wall	LLU	Opposite of J12
J10	M2C8	Tether unit housing	TU	
J11	M2C1	On lever motor	LLU	Additional internal PT100 (green/black–thick)
J12	M2C2	On central controller electronics	LLU	Additional internal PT100 (red/black)
J13	M2C3	On laser	PLC	Additional internal PT100 (orange/brown)
J14	M2C4	On central contr. in the PLC	PLC	Additional internal PT100 (black/red)
J15	M2C5	On PLC motor	PLC	Additional internal PT100 (violett/grey)
J16	M2C6	Control sensor on cooling plate		Underneath PLC between LUs
J17	M3C1	Control sensor on cooling plate		Calibr. sensor PT 100 (inside the shroud)
J18	M3C2	Control sensor on cooling plate		Calibr. sensor PT 1000 (next to J16)

Table 7.1: Position of external PT 100 sensors during first thermal vacuum test

into the chamber, the calibrated PT 1000 next to the J16 chamber reference sensor on the plate, the calibrated PT 100 inside the shroud. For the calibration of the other facility sensors the chamber is evacuated and left to dwell overnight until all the sensors reach an equilibrium and can be calibrated for the tests.

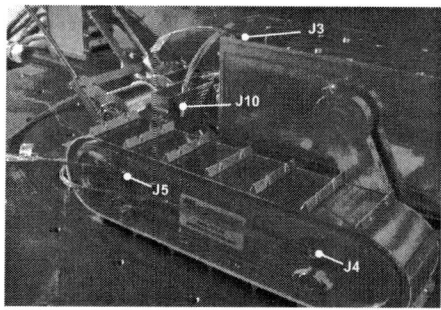

Figure 7.4: Facility sensors on the LOU (right locomotion unit) side of the rover

Figure 7.5: Facility sensors on the LLU (left locomotion unit) side of the rover

Figure 7.6: Facility sensors on the inside of the LLU

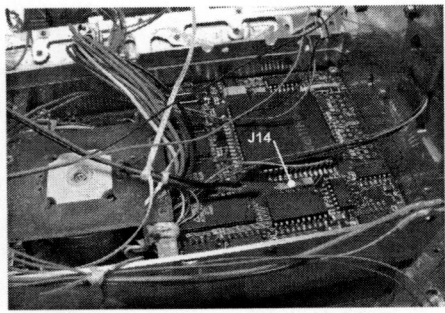

Figure 7.7: Additional PT 100 sensor inside the PLC

Figure 7.8: Additional PT 100 sensor inside the PLC PCB

Figure 7.9: Additional PT 100 sensor inside the LLU

7.3.5.2 Facility thermistors in the second test run

The placement of the thermistors on the rover for the second test is the same as for the first thermal–vacuum test shown in table 7.1, however mounted with different means (see section 7.5). The chamber thermal reference point is on the cooling plate underneath the PLC in between the two LUs. The second calibrated sensor is mounted inside the inner shroud.

7.3.6 Rover thermistors

ID	Location	MRP Node	ADC Channel	Comment
R1	TDU DC–DC switching transistor	TDU	6	
R2	Motor driver IC on motor driver	LLU	3	Locomotion motor
R3	Motor driver IC on motor driver	LLU	4	Lever motor
R4	Locomotion motor	LLU	5	
R5	Motor driver IC	LOU	3	Locomotion Motor
R6	Accelerometer IC on motor contr.	LOU	4	
R7	Accelerometer IC on accelerometer	LOU	5	
R8	Motor Driver IC	PLC	3	Locomotion Motor
R9	PLC Wall – opposite of J1	PLC	5	Reference between rover and facility thermistors

Table 7.2: Position of rover thermistors

The listed rover thermistors (table 7.2) are measured by the rover system while it is in operation. In addition to the temperatures, the rover housekeeping provides information on currents and voltages of the different rover circuits and on the PLC drive angle position by reading out the encoders. The thermistor configuration of the rover is shown in Figure 7.10. Since the temperature measurement depends on the voltage reference level which may drift at low temperatures, the measurement is performed ratiometrically (ADC reference and thermistor reference are the same), so any error in the measurement is minimised.

The rover thermistors are calibrated together with the facility sensors by identifying the offset of the rover thermistors in comparison with the facility calibration sensors after left to dwell.

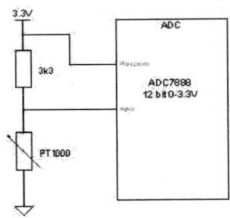

Figure 7.10: Rover thermistor configuration

7.4 Results of first thermal–vacuum test run

The complete test procedure of the first thermal–vacuum test is shown in table 7.3.

From this procedure the problem of the first test setup becomes obvious: Although the system is set to dwell for long periods the cooling plate can not reach the target temperature of $-40°$C. The temperature for radiative exchange is even higher than that of the cooling plate, as the thermal conduction link to the shroud is not optimal. Nevertheless the first test could be conducted and the rover proved its functionality for cold vacuum conditions. Problems with the transmission of the housekeeping data could be observed for the functional test at $-20°$C (step 7), even if the Nanokhod still received the commands sent by the rover interface. The visual control as well as the power consumption drawn from the power supply documented that the rover executed all commands sent to it, however it did not send back continuous housekeeping information. Due to this malfunctioning the test was interrupted and restarted three times, each time letting it run to a later stage of the test programme, with the rover sending out housekeeping data only occasionally. The third attempt was successful with the rover replying reliably with all required housekeeping data and carrying out the complete functional test. Figure 7.11 shows the rover setup in half deployed configuration on the cooling plate inside the thermal chamber and equipped with the thermistors.

The temperatures of all chamber sensors are measured over the complete duration of the thermal–vacuum test cycle. The rover thermistors are only recorded during the functional test, when the rover is delivering housekeeping data.

Step	Description	Action	Notes
1	Installation	Installation of rover; run thermal–vacuum functional test to check rover & test setup, close chamber	
2	Evacuation of chamber	Pump down	
3	Run 20°C thermal–vacuum functional test	Successful thermal–vacuum functional test	T_{plate}=18°C, T_{shroud}=17°C, $P_{chamber}$=2.8×10^{-4} mbar
4	Cool to −10°C	Target temperature set to −20°C, cooling for 7 h, target temperature set to −25°C, cooling/dwelling for 13.5 h	Reference temperature on cooling plate
5	Run −10°C thermal–vacuum functional test	Successful thermal–vacuum functional test	T_{plate}=−10,8°C, T_{shroud}=5.4°C, $P_{chamber}$=1.35×10^{-4} mbar
6	Cool to −20°C	Target temperature set to −50°C, cooling/dwelling for 8 h	
7	Run −20°C thermal–vacuum functional test	Successful thermal–vacuum functional test	Initial problems with receiving housekeeping data; T_{plate}=−22.75°C, T_{shroud}=0.05°C, $P_{chamber}$=1.39×10^{-4} mbar
8	Warm to ambient (20°C)	Warm at maximum gradient +5°C/minute – dwelling for 1 h	
9	Run thermal vacuum functional test	Successful thermal–vacuum functional test	T_{plate}=17°C, T_{shroud}=14°C, $P_{chamber}$=1.66×10^{-4} mbar
10	Reset chamber to atmospheric pressure	Purge chamber with Nitrogen	
11	Run end of test functional test	Successful functional test	Functional test

Table 7.3: Test procedure of first thermal–vacuum test

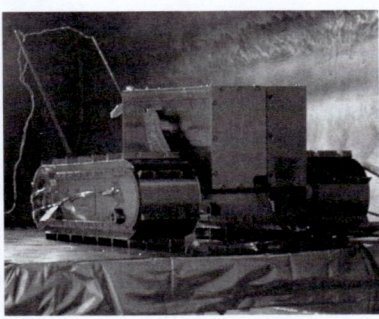

Figure 7.11: Rover setup on the cooling plate for the first thermal–vacuum test

The graphs 7.12, 7.13 and 7.14 show the temperatures of the chamber thermistors during the functional test at $-10°C$ (step 5).

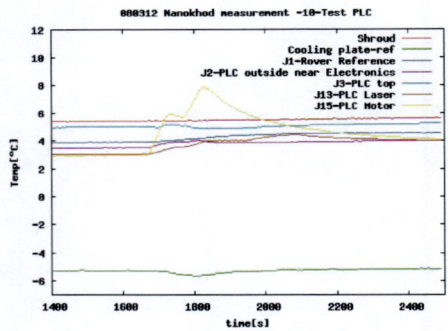

Figure 7.12: Temperature distribution of the PLC during the functional test

The first graph 7.12 shows the temperatures of the thermistors on the PLC components. The temperature difference between the cooling plate ($T_{plate} \approx -5°C$) and the other PLC components shows that the design works well to keep the temperature inside the rover even after relatively long dwell times. In this case this is of course supported by radiation effects caused by the relatively warm shroud ($T_{shroud} \approx +5°C$).

The temperature curve of the PLC motor shows the on and off switching to

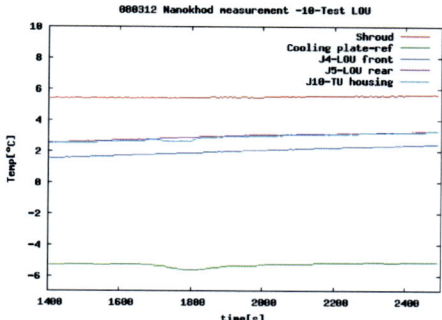

Figure 7.13: Temperature distribution of the LOU and TU during the functional test

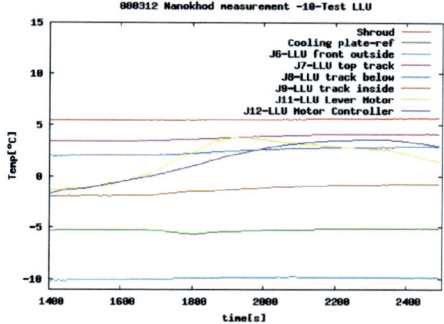

Figure 7.14: Temperature distribution of the LLU during the functional test

move the PLC through an angle of $+90°$ and then $-90°$. The sensor on the PLC wall near the motor electronics follows a similar curve although less extreme and with a slight time delay due to the PCB material and the heat path to the PLC wall.

The laser thermistor also shows a rising temperature. As the laser is only switched on late in the functional test and dissipates much less power, this curve results therefore from the heat radiation of the motor onto the laser thermistor, which is mounted on the laser surface facing the motor.

The cooling plate thermistor shows a decrease in temperature occurring just when the maximum power is dissipated. This results from the thermally controlled cooling circuit which is activated due to the additional heat input in order to keep the plate at a constant level.

The thermistors on the outside wall of the right locomotion unit as well as the one on the tether unit housing in figure 7.13 show a slight temperature increase over the whole test. This results from the heat dissipation of the Tether–Data–Unit and the Tether–Unit PCBs which are the only components which are active throughout the whole functional test.

As expected for the thermistors in the left locomotion unit (figure 7.14), the motor controller PCB and the motor itself follow the activation of the lever movement. The two thermistors on the track foil above and below the motor as well as the two thermistors on the LU side walls near the motor and on the opposite side also show a slight increase in temperatures after a short delay. The thermal heat path to the side walls is realised via the rigid LU yokes. With the motor mounted near the inner side wall it rises here faster than on the outer side wall. The track foils which show only a minimal increase receive the motor heat mainly via thermal radiation.

The track foil thermistor underneath the track reaches an even lower temperature than the thermistor on the cooling plate as it is shielded from the heat radiation by other rover components and the shroud.

7.4.1 First iteration of the thermal network analyses model

For the first iteration of the model the boundary conditions of the rover inside the thermal–vacuum chamber were adjusted to the observed values detected during the first thermal–vacuum test (see section 7.4). The first environmental test showed that the heat conduction of the shroud to the cooling plate was not good enough to achieve the same temperature on both parts. This is why the iterated model was modified to have a separate node and temperature for the shroud and the cooling plate.

This additional node also allowed to model the gap on the backside of the shroud, which was used to observe the rover activities during the first environmental test and thus also respected the radiative impact that the test chamber at room temperature had on the system.

7.4.2 Evaluation of the modelling results with the data of the first test run

For the evaluation and as a preparation for the next model iteration, the calculated temperatures are compared to the measured values. The resulting graphs are discussed in the following paragraphs.

Figure 7.15 shows that for both the LU lever motor and the LLU motor controller PCB, the calculated temperature is more sensitive to the activation of the lever actuator than the measurement indicates. For the motor controller PCB the main reason for the discrepancy is that the modelled PCB assumes the whole board as a single node. This assumes that the dissipated heat is equally distributed over the PCB, resulting in a more distinctive temperature curve. However, the thermistor is mounted in the centre of the PCB, not on the heat dissipating component.

For the motor the calculated temperature is more defined because the thermistor for the measurement is mounted on the motor cradle and thus the temperature change is slower due to the heat path from the motor to the outside of the motor cradle.

Nevertheless the average calculated temperature is higher for both the motor and the PCB. This means the thermal linkage to the locomotion unit walls has to be verified for the next iteration.

The calculated temperature of the central controller PCB shown in figure 7.16 rises more significantly than the measured temperature. The explanation therefore is the placement of the thermistors. The sensor on the central controller is positioned on the face pointing away from both the motor and the PLC controller PCB and its activated heater and is thus completely shielded from the radiation. The heat distribution through the central controller is slow and both smoothes and delays the curve. Nevertheless it seems that the calculated value shows a strong reaction indicating that the conduction values to the board should be reconsidered for the next iteration.

The correlation between the calculated and measured temperature of the PLC motor though is already very good, although the conductive linkage of the motor to its environment could be slightly reduced to optimise the coverage of the outlet ramp of the temperature curve.

The calculated and measured temperatures of the locomotion unit structure (in figure 7.17) match quite well for the general temperature level. The biggest discrepancy is seen from the temperatures on the inside wall of the locomotion lever unit. The measured temperature is lower than the calculated one, due to the radiation influence of the very cold area of the cooling plate underneath

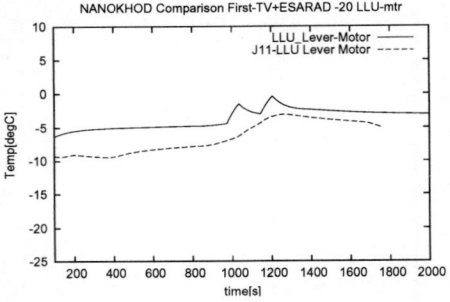

(a) LLU motor (dashed line– measured temperature
(J11), solid line – modelled temperature)

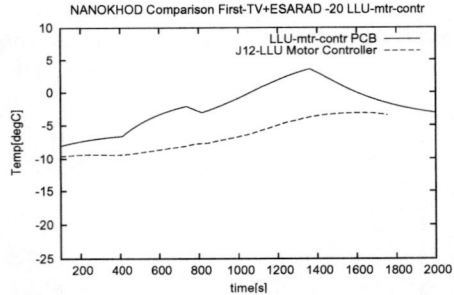

(b) LLU motor controller (dashed line– measured temperature (J12), solid line – modelled temperature)

Figure 7.15: Comparison of the calculated and measured temperatures on the LLU
motor components

the rover directly onto the sensor. There are indications that a reduction of the
conductive linkage of the side walls shall be considered for the next iteration as
the calculated wall temperature responds quite strongly to the heat losses of the
motor.

The temperatures calculated for the track foil with this first model iteration
show a poor agreement with the measured values 7.18. However, this temperature difference not only results from the wrong assumptions and modelling
simplifications within the track foil, which have to be improved in the updated
model, but also from the fact that the sensors are, again, mounted on the sur-

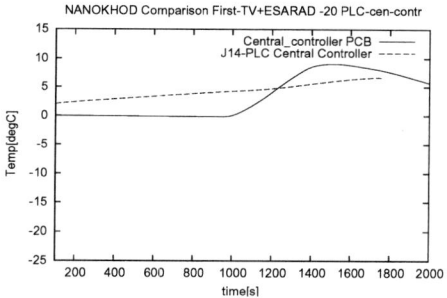

(a) Central controller (dashed line– measured temperature (J14), solid line – modelled temperature)

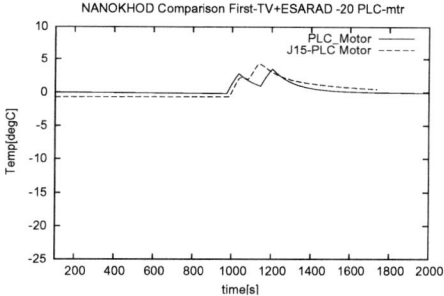

(b) PLC motor (dashed line– measured temperature (J15), solid line – modelled temperature)

Figure 7.16: Comparison of the calculated and measured temperatures on the PLC motor components

face of the components. For the thermistor mounted on the track foil directly above the cooling plate, the radiation effects have a high impact compared to the conductive impact from the track foil; the same is true for the sensor on the upper side of the track foil pointing towards the warm shroud. It is thus necessary to consider not only the adjustment of the thermal nodes, but also the physical mounting of the thermistors (see section 7.5). The correlations of the measured and calculated temperatures must then be reconsidered in the second thermal–vacuum test.

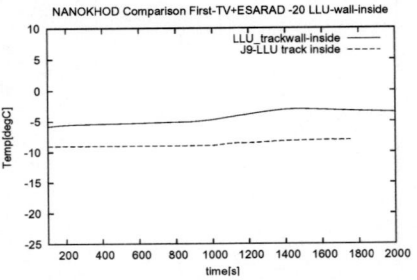

(a) LLU track wall inside (dashed line– measured temperature (J9), solid line – modelled temperature)

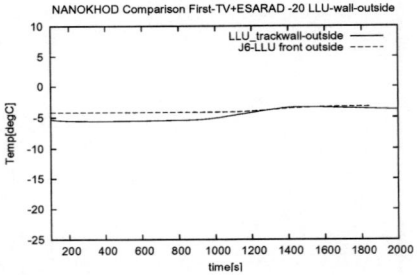

(b) LLU track wall outside (dashed line– measured temperature (J6), solid line – modelled temperature)

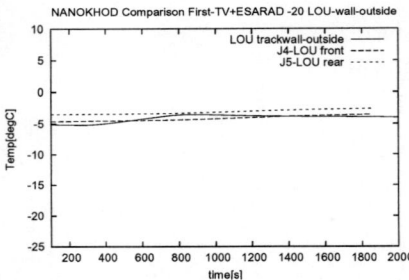

(c) LOU track wall outside (dashed lines– measured temperature (J4 and J5), solid line – modelled temperature)

Figure 7.17: Comparison of the calculated and measured temperatures on the LU side walls

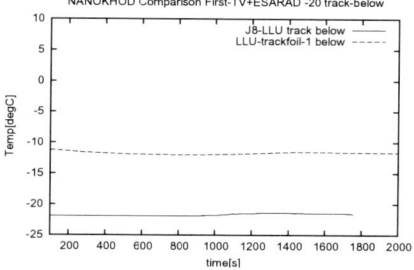

(a) LLU track below (dashed line– measured temperature (J8), solid line – modelled temperature)

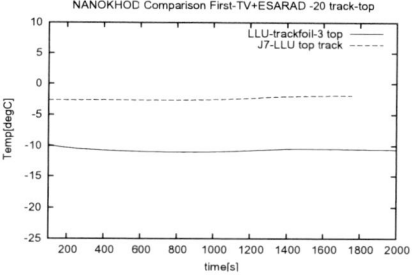

(b) LLU above track (dashed line– measured temperature (J7), solid line – modelled temperature)

Figure 7.18: Comparison of the calculated and measured temperatures on the LU track foils

7.5 Improved mounting methods of the thermistors

Between the two rover TV–tests an investigation was made, which compared different ways of mounting the thermistors. It showed that the method of sensor attachment with the help of a Kapton tape does not provide optimal temperature data. Thus, several methods of mounting the thermistors were tested.

The first tests were conducted without any radiation shield, implying a high impact of the radiation incidence on the sensors. The different configurations of mounting the sensors are described in table 7.4. This test without the radiation shield explicitly shows the low thermal connection of the Kapton tape, see figure 7.19.

Thermistor	Method of mounting
CPT–Silveradhesive	Sensor mounted to cooling plate with two–component glue ECCOBOND 56 C[1] with Catalyst 9
CPT–doublesided tape	Sensor is attached to cooling plate with double–sided tape
CPB	Reference sensor of the heat exchanger, mounted with two–component glue below the cooling plate
CPT–Al–block	Sensor clamped to the cooling plate with an aluminium block screwed to the plate
Kapton tape	Sensor mounted with Kapton tape onto the cooling plate

[1] ECCOBOND 56 C is a silver filled, electrically conductive epoxy adhesive resin. It provides high electrical and thermal conductivity.

Table 7.4: Mounting of the thermistors in the first test

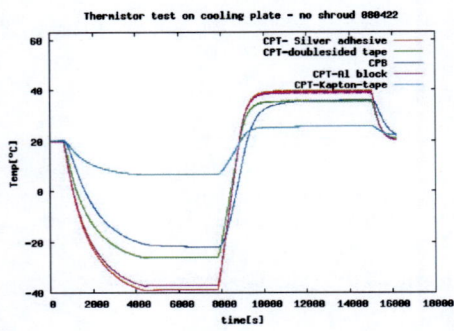

Figure 7.19: First test results from alternative mounting methods – Test setup without radiation shield

In order to provide a good contact between sensor and surface without in-terstices, the best way is to use a two–component glue for the attachment of the sensors. Some of these glues are even enhanced to provide good thermal conductance. The disadvantage of this method is that the sensors are difficult to remove and the risk of damaging the rover surfaces and the Alodine coating is significant.

The test results show that the variety of used tapes can not provide a full contact between the sensor and the surface. Due to low the conductivity of the tape material the heat conductance through the tape itself does not balance the poor contact between the sensor and the surface. Also the effect of radiation becomes visible when comparing transparent and intransparent tapes with each other.

Thermistor	Method of mounting
CPB	Reference sensor of the heat exchanger, mounted with two–component glue below the cooling plate
CPT–Silver adhesive	Sensor mounted on the cooling plate with two–component glue ECCOBOND 56 C with Catalyst 9
CPT–Cu–tape	Sensor mounted with copper adhesive tape
CPT–Al–tape	Sensor mounted with aluminium adhesive tape
CPT–Al–block	Sensor clamped to the cooling plate with an aluminium block screwed to the plate
CPT–doublesided tape	Sensor attached to cooling plate with double–sided tape

Table 7.5: Mounting of the thermistors in the second test

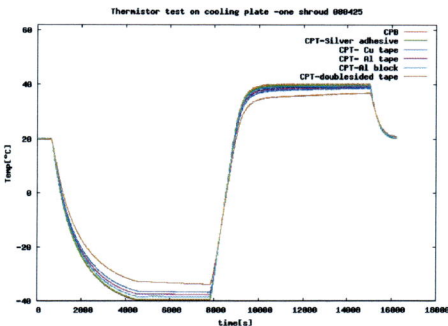

Figure 7.20: Second test results from alternative mounting methods – Test setup with one radiation shield

A second test was conducted with different mounting methods and with one radiation shield reducing the radiation impact. The tested methods of sensor mounting for the second test are listed in table 7.5. The results of the second test in figure 7.20 show that the tested options are already better than the ones from the first test. The effect of radiation becomes obvious when comparing the double–sided adhesive tape for the two cases.

Thermistor	Method of mounting
CPB	Reference sensor of the heat exchanger, mounted with two–component glue below the cooling plate
Inner shroud	Sensor mounted on the inside of inner shroud, opposite from cooling plate
CPT–Silver adhesive	Sensor mounted on the cooling plate with two–component glue ECCOBOND 56 C with Catalyst 9
CPT–Cu–tape–HCP1	Sensor mounted with copper tape and thermal conductance paste – WLP 35
CPT–Cu–tape	Sensor mounted with copper adhesive tape
CPT–Cu–tape–HCG	Sensor glued with two–component heat conductance glue onto copper tape, which is attached to cooling plate surface
CPT–Al–block	Sensor clamped to cooling plate with an aluminium block screwed to the plate
CPT–Cu–tape–HCP2	Sensor mounted with aluminium tape and transparent thermal conductance paste

Table 7.6: Mounting of the thermistors in the third test

In preparation of the second TV–test, the setup with two shrouds was tested in combination with the testing of further sensor mounting alternatives as described in table 7.6.

The temperatures of the third test are shown in figure 7.21 and 7.22. For this test only one cooling cycle down to $-40°C$ was conducted. All mounting methods of this test provide very similar and very good results. Figure 7.22 expands onto the last 2000 s of the measurement. This detailed view shows that there is a small offset between the reference sensor and all other sensors which results from the heat transfer through the cooling plate. The two component silver adhesive, the two copper tapes with heat conductance paste and the heat conductance glue on top of the copper tape show more or less correlating behaviour. The sensor clamped with the aluminium block does not reach the minimum temperature in the given time because of the additional heat capacity from the aluminium block. The worst sensor for this setup is the sensor mounted with copper tape only. This shows that the adhesive tape again does not provide full contact between the sensor and the surface, but obviously it is improved by the use of

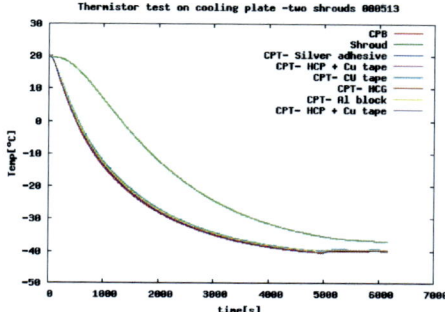

Figure 7.21: Complete measurement of the third thermistor test – Test setup with two radiation shields

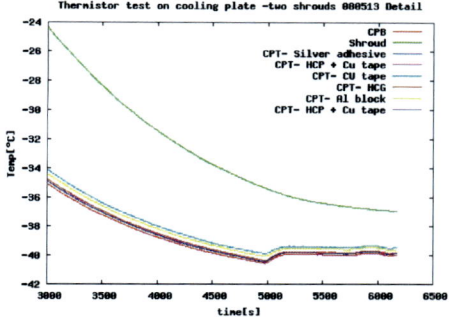

Figure 7.22: Focus on the nearly steady state temperatures of the third thermistor test

heat conductance paste.

In order to not damage the Alodine coating of the rover surface, it was thus decided to use copper tape with heat conductance paste for the mounting of the thermistors in the second thermal–vacuum test. The mounting is shown in figure 7.23.

Figure 7.23: Mounting of Thermistor with copper tape and heat conductance paste between thermistor and rover surface

7.6 Results for second thermal–vacuum test run

Step	Description	Action	Notes
1	Installation	Installation of rover, run thermal–vacuum functional test to check rover & test setup, close chamber	
2	Pump Down	Evacuation of chamber	
3	Run 20°C thermal–vacuum functional test	Successful thermal–vacuum functional test	T_{plate}=18.2°C, T_{shroud}=18.2°C
4	Cool to −10°C	Target temperature set to −15°C, cooling/dwelling for 2 h	
5	Run −10°C thermal–vacuum functional test	Successful thermal–vacuum functional test	T_{plate}=−14.3°C, T_{shroud}=−8.5°C
6	Cool to −20°C	Target temperature set to −25°C, cooling/dwelling for 2 h	
7	Run −20°C thermal–vacuum functional test	Successful thermal–vacuum functional test	T_{plate}=−23.7°C, T_{shroud}=−17.5°C
8	Cool to −30°C	Target temperature set to −35°C, cooling/dwelling for 2 h	

9	Run $-30°$C thermal–vacuum functional test	Successful thermal–vacuum functional test	Initial problems with receiving housekeeping data; $T_{plate}=-32.8°$C, $T_{shroud}=-25.2°$C
10	Warm to ambient $(20°$C)	Warm at maximum gradient $+5°$C/minute – dwelling for 1 h	
11	Run thermal–vacuum functional test	Successful thermal–vacuum functional test	$T_{plate}=15.4°$C, $T_{shroud}=11.1°$C
12	Reset chamber to atmospheric pressure	Purge chamber with Nitrogen	
13	Run end of test functional test	Successful functional test	

Table 7.7: Test procedure for second thermal–vacuum test

The test procedure for the second thermal–vacuum test is shown in table 7.7. The rover setup for this test (figure 7.24) is the same as for the first one (described in section 7.4), except that the cable guides in the rear are fixed at a slightly steeper angle to fit them within the inner shroud.

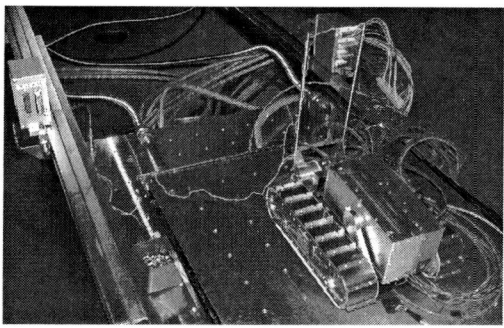

Figure 7.24: Rover setup on the cooling plate for the second thermal–vacuum test

The initial problems to receive housekeeping data occurring on step 7 of the first thermal–vacuum test (Table 7.3) and on step 9 for the second test (Table 7.7) showed the first temperature effects on the functioning of the electrical components. During both tests, the signal transfer to the rover, as well as the circuits on the rover itself worked reliably for all temperature levels. However the return signal with the housekeeping information initially failed to be transferred to the rover interface for the mentioned steps. This occurrence was evident from the reliable operation of the drives, the LEDs and the laser, while the housekeeping signals were only received irregularly. For both tests, the test sequence was interrupted for step 7 respectively 9 and restarted twice from the beginning, before the housekeeping data were received dependably for the whole test sequence.

The instability of the signal transmission from the rover is caused by a timing problem of the transmission electronics in the tether unit of the rover. The temperature affects the timing of the differential driver and the pulse shaper, which is providing pulses to the transformer for the transmission on the tether. At low temperatures the signal rise of the differential driver is slowed down, so the pulses are to low for the transformer and fail to be transferred.

For the application of the rover at lower temperatures the transmission signals and their timing need to be adjusted, so all data is transmitted reliably back to the rover interface on the lander side.

For the second thermal–vacuum test the measured temperatures of step 7 with the environmental temperature of $-20°$C are discussed specifically. Generally, the behaviour of the temperature curves in the second thermal–vacuum test is similar to those of the first test results. The main differences can be seen in the rise and duration of the lever motor component temperatures. In the second TV–test, the articulation of the PLC motors were driven for a longer time, rotating the PLC 4 times by $\pm90°$, see section 7.3.4. The graphs reflect this longer rise time of the temperatures. The dwell times were slightly shorter for this test. However, again there is a temperature difference between the rover and the cooling plate, although this time the shroud temperature was below the rover temperatures. This demonstrates the performance of the thermal design, which keeps the heat within the rover.

Figure 7.25 shows the temperatures of the PLC components during the functional test at environmental conditions of approximately $-20°$C. The thermistor on top of the PLC (J3) shows an approximately constant temperature for the whole functional test. The thermistor on the PLC motor (J15) shows a clear temperature rise, when the motor is activated. Even the short phases when the motor is switched off before changing the direction of rotation becomes obvious in the graph. The thermistors on the PLC side wall, the one near the PLC drive electronics (J2), the PLC reference point (J1) and the one on the laser (J13) all show a slightly increasing temperature during the motor activity: The thermistor near the drive electronics reacts to the dissipation by the heater, the sensor on the

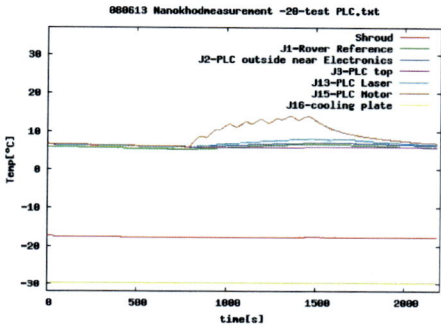

Figure 7.25: Temperatures of the PLC for the functional test at $-20°C$

laser is in the direct field of view of the heat dissipating motor. The same is true for the part of the PLC side wall onto which the reference sensor is mounted.

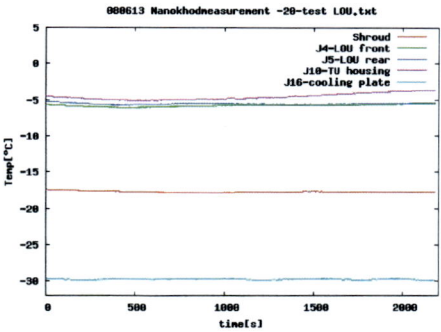

Figure 7.26: Temperatures of the LOU and TU for the functional test at $-20°C$

The temperatures in the LOU also behave similarly to the first test results, see figure 7.26. The thermistor on the tether unit (J10) as well as the two sensors on the LU side wall (J4 and J5) show slowly rising temperatures during the whole test, with the PCB in the TU and the tether data unit PCB in the LU dissipating heat during the rover activity of the functional test.

For the LLU most of the thermistors respond to the activation of the lever

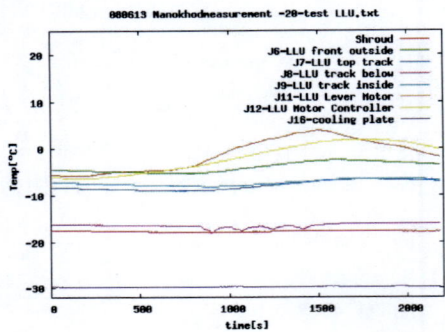

Figure 7.27: Temperatures of the LLU for the functional test at −20°C

motor, see figure 7.27. The temperatures of the motor controller (J12) and the motor (J11) rise during the PLC rotation movement. This time the switching of the motor can not be seen as clearly as for the PLC motor which is due to the mounting of the thermistor on the outside of the motor cradle which dampens the temperature curve. Due to the low thermal conductivity of the PCB–material, which slows the effect, the motor controller curve rises slower but also needs more time to cool down after the motor is switched off. The thermistors on the side wall (J6 and J9) as well as the ones on the track foil (J7 and J8), all mounted near the activated drive, have curves that slowly rise when the PLC moves. The sensor temperature on the outside wall (J6) increases the most with the lever motor being mounted to the outer side wall. The wave shaped behaviour of the track foil temperature below the locomotion unit (J8) can be explained with the rotation of the gear components with hotter and colder parts facing the track foil.

7.6.1 Evaluation of the modelling results with the test data of the second test

In order to evaluate the thermal model, it is modified to reflect the boundary conditions of the second test with a completely closed shroud covering the rover. For a representative comparison of the calculated and measured temperatures the functional test of step 7 was chosen.

The test results with the more uniform radiation environment matches the calculated results more exactly. The comparison of the LLU motor and the LLU

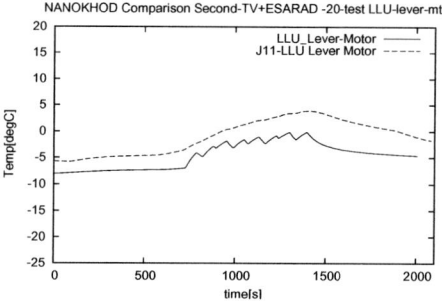

(a) LLU motor (dashed line– measured temperature (J11), solid line – modelled temperature)

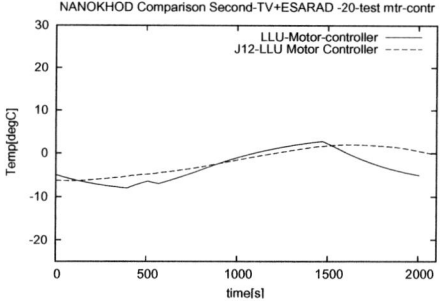

(b) LLU motor controller (dashed line– measured temperature (J12), solid line – modelled temperature)

Figure 7.28: Comparison of the calculated results (iterated model) and temperatures of the second thermal–vacuum test for the LLU motor components

motor controller in the graphs of figure 7.28 show a good correlation of the temperature levels. For the motor the calculated temperature shows again a more defined profile for the heat dissipating phase of the PLC activity. However, this can be explained with the position of the sensor, which is on the outside surface of the motor cradle and not directly on the motor itself. The smoother curve thus results from the heat path through the cradle and the inherent slowdown effect. The motor controller temperatures correlate quite well, even if the heat capacity and the thermal conductance of the motor controller have to be reconsidered as the calculated temperature drops much faster than the measured value.

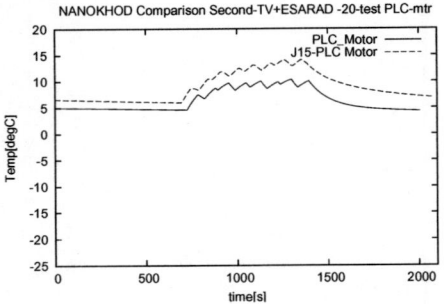

(a) PLC motor (dashed line– measured temperature (J15), solid line – modelled temperature)

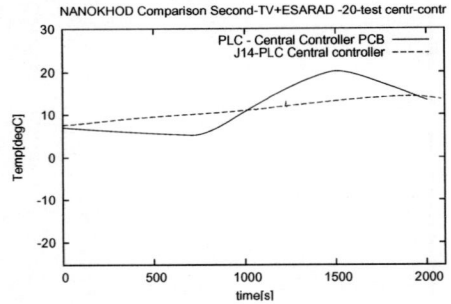

(b) Central controller (dashed line– measured temperature (J14), solid line – modelled temperature)

Figure 7.29: Comparison of the calculated results (iterated model) and temperatures of the second thermal–vacuum test for the PLC motor components

For the PLC motor the measured and calculated data (figure 7.29) show the same behaviour, however the calculated data is lower, which indicates that the losses from the motor to surrounding parts are slightly too high and need to be reconsidered in the model's next iteration. The comparison of the central controller PCB data shows the same behaviour as for the first test. For a better investigation of this component it shall be divided into more nodes in the next iteration of the model.

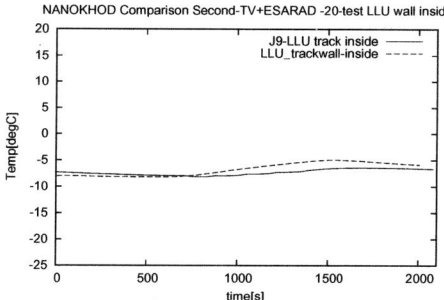

(a) LLU track wall inside (dashed line– measured temperature (J9), solid line – modelled temperature)

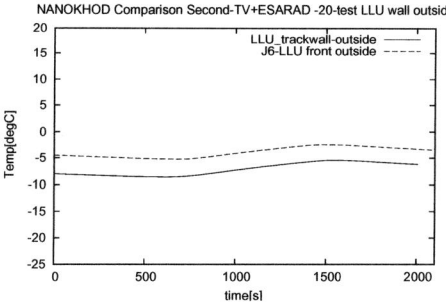

(b) LLU track wall outside (dashed line– measured temperature (J6), solid line – modelled temperature)

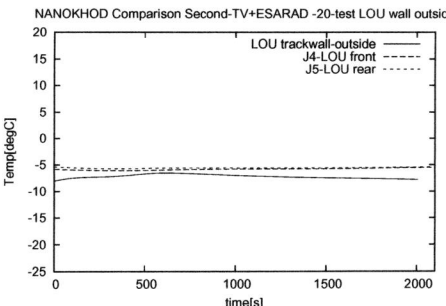

(c) LOU track wall outside (dashed lines– measured temperature (J4 and J5), solid line – modelled temperature)

Figure 7.30: Comparison of the calculated results (iterated model) and temperatures of the second thermal–vacuum test for locomotion unit side walls

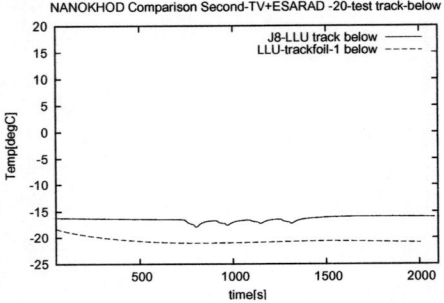

(a) LLU track below (dashed line– measured temperature (J8), solid line – modelled temperature)

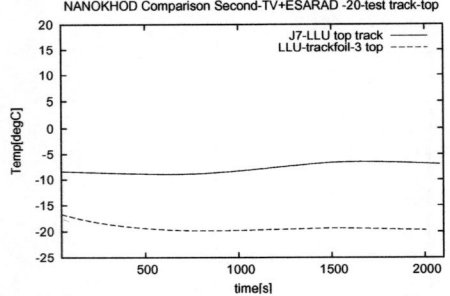

(b) LLU above track (dashed line– measured temperature (J7), solid line – modelled temperature)

Figure 7.31: Comparison of the calculated results (iterated model) and temperatures of the second thermal–vacuum test for the LU track foils

The data of the thermistors on the locomotion unit walls of the second TV–test show the influence of the radiation environment when comparing the data with the calculated values in figure 7.30. The temperature curves for all thermistor positions correlated quite well. However the measured and calculated temperature levels do not completely match.

This mismatch may be a dependency of the track foil data, see figure 7.31. The curves for the track foil show a large mismatch. In the next iteration of the thermal model the track foil modelling and the origin of this discrepancy needs to be investigated further.

7.6.2 Optimised and validated model

Based on the previous results and observations the thermal model was iterated, to achieve a better correlation between the measured and the calculated data. For this iterative step the following changes were implemented:

- Adjustment of the thermal conductance data, the conductive interfaces and radiation impacts of the trackfoil.
- The heat capacity of the Laser was reduced, as it was approximated in the first model with an aluminium cylinder of a too high mass.
- Adjustment of the conductance and the conductive interfaces of the rover drives to the supports.
- Adjustment of the conductance and the conductive interfaces of the PCBs, especially of the standoffs supporting the central controller PCB and the LLU motor controller.
- Adjustment of the conductance and the conductive interfaces of the PLC walls.
- Separation of the PLC controller PCB into the small section with the heat dissipating drive electronics and the rest of the PCB.
- Introduction of conductors with different contact heat transfer coefficients, varying with the contact pressure, based on the sensitivity analyses (see section 6.4.2).

7.6.3 Comparison of test results and iterated thermal modelling results

When comparing the same component temperatures as in section 7.6.1 with the results of the third iteration of the model, the graphs in this section show a better correlation. However the temperature curves of the calculated and the measured data do not match by 100%, due to simplifications made within the model as well as effects of the temperature measurements. In general, the simplification lead to the effect that the calculated temperatures provide an average value for the whole component. This effect is even greater for components, which partially dissipate heat like the PCBs, with the average taken over areas which provide a higher thermal gradient. Also big components show a higher inaccuracy as the average temperature is taken over a larger volume.

Systematic measurement errors result from the fact, that the temperatures are only determined on one surface of the component and thus depend on the conductance within the component itself. Thus the measured value will give only local information. Additionally the mounting of the thermistors on the surfaces causes an impact of radiation effects from the surrounding components. One of the sensor surfaces will always be influenced by the radiation, while the component itself distributes this impact into its inside without any contact conductance.

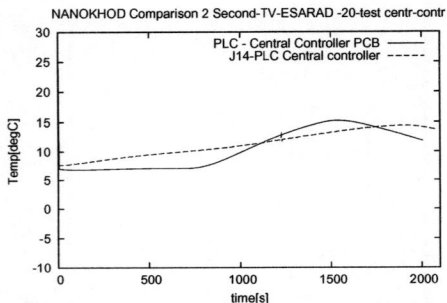

(a) PLC central controller (dashed line– measured temperature (J14), solid line – modelled temperature)

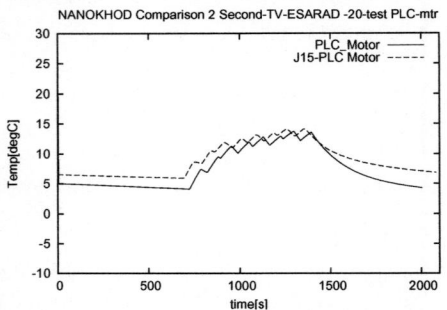

(b) PLC motor (dashed line– measured temperature (J15), solid line – modelled temperature)

Figure 7.32: Comparison of the validated model results and temperatures of the second thermal–vacuum test for the PLC motor components

The comparison of the measured and calculated temperature data of the central controller PCB (figure 7.32(a)) shows a slightly different profile. The calculated curve rises slowly to a maximum during the whole rover functional test, while the measured temperature stays constant until the PLC articulation drive starts moving. After the activation the calculated value rises with a higher gradient than the measured curve until the motor is deactivated. The reason for this is that the sensor of the central controller is mounted on the surface pointing away from the PLC controller which is dissipating heat as soon as the rover functional test is initiated. The sensor is in the centre of the central controller PCB, thus as far away from the standoffs towards the PLC controller as possible. This means the heat resistance from the heat dissipating PLC controller to the sensor is much higher than to other parts of the board. Thus the average temperature of the board is higher than the measured temperature. Opposing to this, the dissipated heat of the motor is radiating directly on the surface of the central controller and thus the sensor. This is why the sensor is showing a much stronger reaction when the motor is switched on, while the calculated temperature average of the board is rising more slowly.

The curves showing the calculated and the measured temperatures of the PLC articulation motor show a well matching profile (figure 7.32(b)). The sensor is mounted directly on the motor clamp and thus the correlation of the data is very good.

Figure 7.33(b) shows a good compliance of the calculated and measured temperature of the motor controller inside the left locomotion unit. The calculated temperature again shows a distinctive curve, while the measured curve is following the same trend with a smoother line. The sensor is measuring a smoother temperature behaviour as it is again on the surface of the motor controller PCB, and thus the dissipated heat which is reaching the sensor is delayed by the heat path through the board. For the calculation the heat is dissipated directly within the component and thus the temperature curve follows the activation of the motor controller without that delay.

The articulation motor in the LU (figure 7.33(a)) shows a similar graph as the PLC motor above. The calculated temperature matches nicely the measured temperature curve; however, the measured data is less distinctive due to the position of the sensor. The sensor is mounted on the outside of the motor cradle, where the maximum thermal resistance of the part applies. Opposed to that the calculation assumes the average of the thermal resistance from zero on the inside of the motor to the maximum on the surface of the cradle. At the same time the thermal radiation effects within the locomotion unit are stronger due to the colder parts surrounding the motor. Thermal radiation from the motor controller

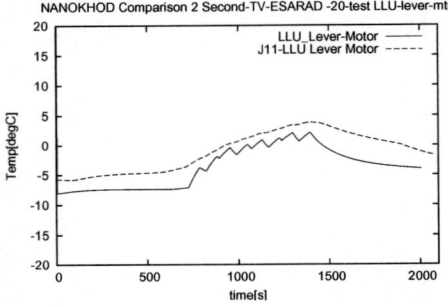

(a) LLU motor (dashed line– measured temperature (J11), solid line – modelled temperature)

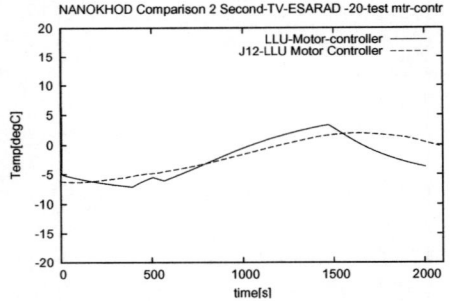

(b) LLU motor controller (dashed line– measured temperature (J12), solid line – modelled temperature)

Figure 7.33: Comparison of the validated model results and temperatures of the second thermal–vacuum test for the LLU motor components

also explains the slightly rising temperature value of the motor due to radiation effects within the locomotion unit before the articulation drive is activated.

The two temperature curves of the track wall (figure 7.34(a)) show the same trend. However the calculated temperature rises much more after the activation of the articulation drive. The explanation for this is that the thermal model does for example not consider the shielding by the harness inside the LU. It thus assumes much higher radiation effects from the motor on the inner LU side wall.

The graph 7.34(b) of the track wall on the outside shows almost the same curve for the calculation and the measurement. However, the measured curve is

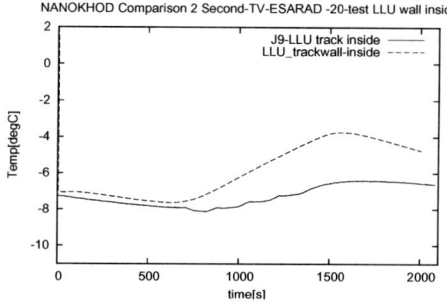

(a) LLU track wall inside (dashed line– measured temperature (J9), solid line – modelled temperature)

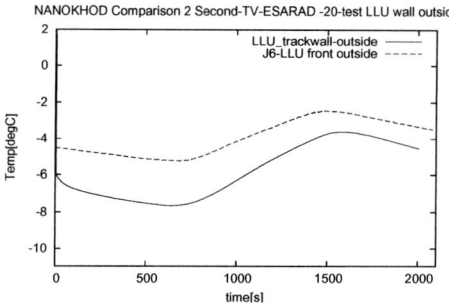

(b) LLU track wall outside (dashed line– measured temperature (J6), solid line – modelled temperature)

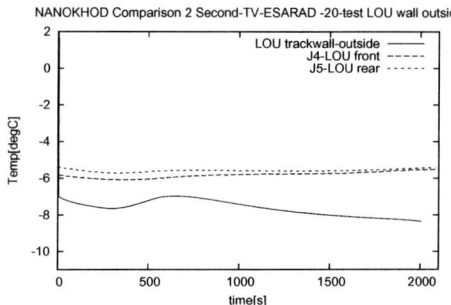

(c) LOU track wall outside (dashed lines– measured temperature (J4 and J5), solid line – modelled temperature)

Figure 7.34: Comparison of the validated model results and temperatures of the second thermal–vacuum test for locomotion unit side walls

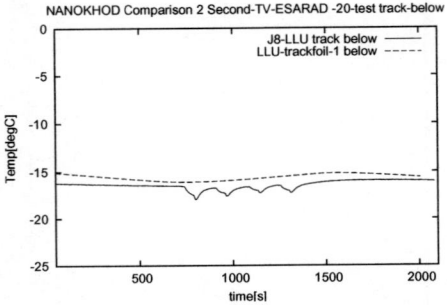

(a) LLU track below (dashed line– measured temperature (J8), solid line – modelled temperature)

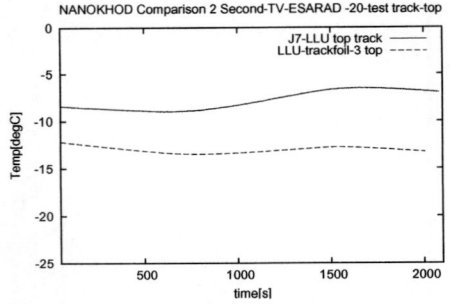

(b) LLU above track (dashed line– measured temperature (J7), solid line – modelled temperature)

Figure 7.35: Comparison of the validated model results and temperatures of the second thermal–vacuum test for the LU track foils

always ~ 1°C higher than the calculated curve. As the thermal model assumes an average temperature for the cooling plate, it does not take into account the thermal gradient between the shielded temperature underneath the rover and of the area around the rover, which receives the radiation from the slightly warmer shroud. Due to that the calculated temperature is assuming a lower radiation temperature of the cooling plate and thus results into a lower level temperature on the side wall.

For the outer side wall of the right locomotion unit (figure 7.34(c)) both effects can be seen, the shielding of the cabling as for the left inner side wall and the assumption of a lower radiation temperature of the cooling plate. Due to that the calculated temperature shows a clear increase of the temperature for the time the motor heaters are switched on. The effect of the activated heater in the LOU in the beginning of the functional test is only a marginal increase of the measured curve shown by the sensor on the rear of the locomotion unit. The effect is very small as the heater on the motor driver is shielded from the outside LU side wall.

Figure 7.35(a) shows that the curves for the lower side of the track foil match quite nicely. The only differences are some waves in the measured curve during the activation of the articulation drive. These result from the rotation of the drive providing a changed radiation interface to the sensor, by moving a colder surface of the drive towards and then away from the sensor.

For the upper side of the drive a similar effect as for the side walls can be observed. The calculated and measured curve match very well (Figure 7.35(b)), however the measured curve is always higher than the calculated one. This is again due to the use of an average temperature value for the shroud and the cooling plate within the calculation.

7.7 Influence on the thermal system design

Although there are still some discrepancies between the model and the measured temperatures it can be stated that they are generally insignificant and can be explained by the applied simplifications and methods. By keeping these influences in mind, the iterated model can now be used to provide information about the temperature distribution of all rover components which could not be measured due to the limited amount of thermistors or limited accessibility. Additionally the model can be used to implement different boundary conditions, in order to analyse the thermal behaviour of the rover within any selected environment. It can be used to disclose critical areas for the applied environmental conditions and allows the optimisation of the thermal system concept if necessary. With these results the thermal system of the rover can be validated for the target environment of any mission.

7.7.1 Thermal network analyses model – Mercury environmental conditions

The last iteration of the model implements the boundary conditions as expected for the Mercury lander of the BepiColombo mission scenario.

The surface conditions are modelled with a much larger surface than the rover dimensions covered by a half dome. The plate is assigned with a surface temperature of $-180°C$ and a thermal conductance for regolith of $\lambda = 0.01$ W/mK [61]. The deep space temperature of the half dome above the rover is set to 3 K.

The surface operation of the Nanokhod rover for the first 48 h, which is implemented in the model calculation is given in table 7.8. This surface operation sequence for the Nanokhod is based on the BepiColombo operational scenario.

Start Time	Duration	Activity	$T_{surface}$ (°C)	Rover modes
00:00:00	08:01:10	Checkout Sequence	−150	Localisation, Science APXS, Science MIMOS, Science MIROCAM
08:01:10	00:30:17	Complete Path (Lander Egress)	−150	Locomotion, PLC Move, Localisation, Locomotion, Localisation
08:31:27	08:09:23	Measurement Sequence	−180	Locomotion, PLC Move, Science MIROCAM, PLC Move, Science APXS, PLC Move, Science MIMOS, PLC Move
16:40:51	07:19:09	Rover Off	−180	Idle
24:00:00	00:30:17	Complete Path Segment 1	−180	see above
24:30:17	00:30:17	Complete Path Segment 2	−180	see above
25:00:35	00:30:17	Complete Path Segment 3	−180	see above
25:30:52	08:09:23	Measurement	−180	see above
33:40:15	14:19:44	Rover Off	−180	Idle

Table 7.8: Nanokhod Mercury surface operation – first 48h

The temperature distribution at two specific points in time of this operation procedure are considered in more detail. Figure 7.36 shows the rover temperatures

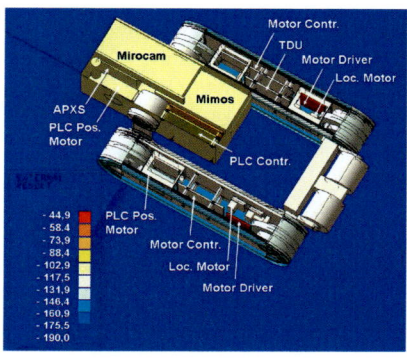 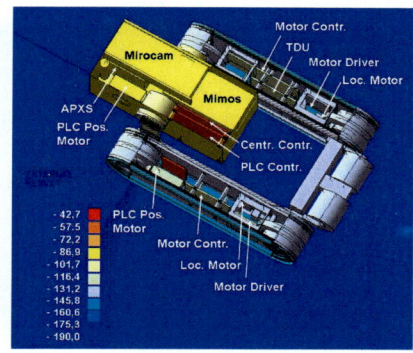

(a) Rover after locomotion   (b) Rover after science cycle

Figure 7.36: Temperatures in the rover after science cycle and locomotion activity

after the first *Complete Path Segement* activity (figure 7.36(a)) and respectively after the *Measurement Sequence* activity (Figure 7.36(b)). In both cases the temperature within the rover varies between approx. $-44°$C on the hottest component and $-180°$C on the Mercury surface node.

During the scientific measurements the highest temperature can be observed on the central controller PCB within the PLC. Also the PLC motor is quite warm due to the repetitive PLC rotation which is necessary for the positioning of the instruments. As the last active instrument of the measurement cycle the MIMOS instrument is still on a temperature level of approx. $-70°$C.

For the locomotion activity mainly two areas reach an increased temperature: The locomotion driver PCBs in the rear of each locomotion unit and slightly less the central controller PCB in the PLC. With the low heat conduction properties of the PCB material the dissipated heat of the electrical components is only transferred slowly to other parts of the rover.

These two graphs only show two examples for thermal distributions within the system for any rover mode, that can be provided by the model. By applying all operational modes for the planned duration the thermal extremes can be assigned, to which each components is exposed during the changes between idle and operational phases. The definition of these boundary temperatures supports any future design steps.

They also help to define the conditions provided for the scientific instruments.

If it becomes necessary, measures can be taken to allocate the required boundary conditions for the instruments. Mounting devices, materials or surface conditions can be adjusted; even small heaters could be included in the design to ensure that the instruments can be used within their operational temperature range.

8 Conclusion

The development of the miniaturised robotic space system Nanokhod, as described in this thesis, has made a big step forward in direction of a flight model rover. The mechanical design of the Nanokhod rover was developed from a conceptual stage to an engineering model which is able to withstand the demanding conditions of a Mercury mission. The electrical design was completely revised to match with the severe power limitations of a night side Mercury lander. The new rover design was thus optimised to meet the central mission criteria of mass, volume and power in order to prepare the rover for a flexible implementation on a variety of missions. The design that has been achieved represents a highly integrated, mobile payload system for the exploration of planetary surfaces.

8.1 Nanokhod for the Mercury environment

The aim of the project was to build a rover which is compatible to a mission scenario that targets for surface exploration on the night side of Mercury. The transfer to the Mercury surface implies severe constraints on mass, volume and power of the system and exposes it to a high number of thermal cycles and a severe landing shock when impacting on the surface. For the operation on the surface the rover system has to cope with extremely low temperatures combined with a high vacuum and an abrasive regolith surface. Despite these demanding requirements and limited financial resources a model was manufactured that successfully matches this mission profile.

The original intention was to submit the rover model to a full set of environmental tests. However delays created due to problems with the drive system blocking and eventually led to the decision to cancel the tests. The drive blockage problem was investigated and it was concluded that temporary rises of the torque load on the stepper motor causing it to miss a step and thus to stall. As the project proceeded, the problem occurred less frequently with the duration of the drive operation going up. Nevertheless several tests and analyses were conducted to investigate the torque output of the gear stages in order to fully understand all related issues. The solution found was to improve the tolerance control and to select an alternative motor, which allows for a higher margin.

Despite this motor issue the rover has reached an advanced state of development, particularly because a clear roadmap has been elaborated on how to fully solve the problems that have been identified and thus to provide a reliable solution. Although, currently no mission opportunities are in sight, the rover is expected to be seriously considered for future missions to the moon and other destinations, which only provide limited resources and still aim for mobility.

8.1.1 Rover design

The rover design task was initiated by reviewing the requirements of previous related projects , in accordance with the BepiColombo mission definition. The requirements for the rover application on the cold night side of the Mercury were the baseline for this rover development.

In compliance with these requirements, the first step was the design of the rover mechanical and electrical components based on previous Nanokhod rover developments. The design of the rover had to provide a number of enhanced features for the Mercury mission:

- The overall design had to be compatible to temperatures down to $-180°$C.
- The drive system required a more durable and precise solution, which includes a dry lubrication treatment.
- The tether unit was lifted to prevent a *bulldozing* effect on the regolith surface.
- Deployable tether guides were implemented to fully fit into the allowable volume.
- The locomotion units were stabilised by internal yoke units to provide sufficient strength and rigidity to withstand the high landing shock load.
- A concept design of a lander hold–down–device (HDD) was devised which secures the rover during launch and landing, Additional locking devices required by the HDD were implemented in the rover design.
- The track seal had to be redesigned for the required temperature range to prevent regolith from entering the locomotion units.
- The rover control partition and the P/L partition of the PLC had to be clearly separated in order to provide the necessary volume for the GIPF integration.
- A thermal model had to be developed to validate the thermal design and

the thermal control concept of the current rover model and to provide an essential basis for future design steps or for alternative missions.

- An optimised electronic subsystem had to be developed for the low temperature environment.

8.1.2 Outcome of the development

The result of this development process is a rover design which is compatible to the Mercury environmental conditions. Although the current model was not yet tested to the extreme environmental conditions as initially intended, the conducted tests provide a design validation and show that the rover design is a promising solution for dustbox tests and for TV tests which cover the complete thermal range.

Parameter	Unit	Require-ment	Manu-factured	Notes
Stowed length	mm	240	232	
Stowed width	mm	165	162.2	
Stowed height	mm	65	67.4	Extra height due to protruding LEDs, which are raised for a better visibility
Rover mass	g	1800	1820	Manufactured mass includes the central subsystem PCB which is not included in the requirement
Instrument mass + central subsystem	g	900	550	From GIPF project and measured values – this does not include the central subsystem mass

Table 8.1: Odometry matrix of manufactured model

The verification of the rover odometry (table 8.1) and the completed functional tests show that the design meets all functional requirements except for the required maximum speed (table 8.2). However this non–conformance is acceptable as there is a reasonable margin to cover the increase of speed to the required value. In addition, the rover is capable to cover the required driving distances in the given time even at the lower speed. Therefore, the limitation of the rover's maximum speed has no negative impact on the overall mission.

Requirement	Description	Status	Notes
Mobility – reverse	Reverse by at least its own length	Fulfilled	
Mobility – spot turning	The rover shall perform a spot turn $\geq 90°$	Fulfilled	
Mobility – obstacle	Climb a step of 0.1 m or traverse a ditch of 0.1 m width	Fulfilled	
Mobility – speed	Maximum traverse speed of 5 m/hr	3.06 m/h	Limited by planetary gearhead
Payload Cab positioning	Orientation of PLC apertures to a surface at inclinations from vertical to horizontal	Fulfilled	
Payload Cab contact	The rover shall push the payload cab against a hard surface with a force of ≥ 1 N	Fulfilled	
Rover – peak power	The rover shall consume not more than 6 W peak power from a 28 V source	Fulfilled	

Table 8.2: Requirements matrix for the rover functional test

After the successful functional tests in atmosphere and room temperature conditions, the first environmental tests were conducted in a thermal vacuum chamber. Although the tests could not be taken to the extreme limit of down to $-180°$C the tests validated the functionality of the design for temperatures down to $-40°$C and vacuum conditions. This is a first qualification step and the results are promising for the full TV–tests.

For the extreme environmental conditions and also in preparation of the full TV tests the development of a more sophisticated thermal mathematical model became necessary. The model was validated with the data of the above–mentioned thermal vacuum tests. The validated model can now be used to predict temperature related behaviour of specific components and thus decrease the risk of mistakes in the choice of components and facilitating the preparation of next design steps. The thermal model additionally allows for an easy revision of the thermal design, e. g. for alternative mission scenarios.

8.2 Further Development

Although the current rover design meets all requirements, there are still some critical issues which have to be addressed in order to progress to a full flight level of implementation.

For the electronic design the MRP node functionality should be implemented as an ASIC, which is a critical step in the FM development in terms of cost, risk and schedule. The use of an ASIC processor is essential as this solution provides the ultimate in miniaturisation and guaranteed radiation performance. An alternative solution is the use of an FPGA device which combines analogue and digital functionality. However such a device still has to be approved for space applications.

Even though the current design provides the necessary volume for the scientific instrument heads of the geochemical instrument facility (GIPF), further work needs to be done to fully integrate the instruments electronics with the rover electronics. A baseline concept for the integrated electronics was already developed within the HICoPS study [44]. This concept needs further evaluation and breadboarding activities, before being integrated in the Nanokhod rover.

On the mechanical side the solution of a reliable drive implementation was analysed extensively within this thesis. However more research is needed on that issue. For the next development step it is essential to qualify the drive components separately from the rover on an equivalent test stand which allows for a detailed performance measurement. This will ensure that all contributing factors are fully identified. In addition to the experiences gained in the drive development and testing which have been carried out so far, separate testing of all components will lead to a reliable drive unit solution. This solution should also include the development of an encoder for a brushless DC motor which is applicable for the environmental conditions.

For the integration of electronic parts suitable for space applications further component testing need to be carried out in all future development steps. As no components are available which are space–qualified for these extremely low temperatures, standard components have to be chosen and a full component characterisation has to be conducted for the flight model. This is a labour intensive task which implies that the number of different components should be minimised. Furthermore, components and manufacturing processes have to be chosen that are likely to work at low temperatures.

8.3 Nanokhod development plan and ROM costing

A development plan for a Nanokhod flight model will involve three major steps:

Environmental testing: The first step towards the flight model development is the continuation and completion of the MRP environmental testing. This offers the possibility to gain as much information as possibility to prepare the EQM redesign.

EQM Development: The EQM development includes the design of the flight model, the development and manufacture of a desktop model for electronics and for the drive unit as well as the design, the manufacture, and the qualification of the EQM model.

Flight Model & Flight Spare: This development phase comprises the manufacture and qualification of the Flight and Flight spare models.

The schedule for the suggested approach is given in appendix D and the ROM costs are given in table 8.3.

	Man power [k€]	Other items [k€]
MRP environmental test	68	25
EQM activity	2160	1600
FM and FS activity	808	470
Total	3036	2095

Table 8.3: Nanokhod costing for flight design development

8.4 Alternative targets for the rover

A mission to the Mercury surface is currently unlikely after ESA cancelled the lander of the BepiColombo mission. Still, there are several worthwhile alternative targets for the Nanokhod. The Nanokhod rover system prepared for the mission conditions of the Mercury environment is a perfect baseline for a variety of different targets and missions requiring local mobility. Of course, new scientific objectives will require a certain degree of redesign to adjust the rover to the different environmental needs; however the high level technical development that has been achieved offers a solutions at a minimal risk.

Recently, the Moon has become very popular in consideration of proposed visions. First studies provide evidence that the rover is a suitable scientific payload for a small satellite mission to the Moon (see appendix H and [62]).The night side environmental conditions of the Moon are similar to Mercury and would allow an easy adaptation of the current concept. For a dayside landing, attention would need to be given to the thermal design of the Nanokhod model. However, with the developed thermal model that has been described above, this task is already perfectly primed.

Although the planet Mars is currently visited by systems with a higher level of mobility (MER and the upcoming missions ExoMars and MSL), it is also an interesting target option for a Nanokhod rover system, especially if mission scenarios in the more demanding polar regions are considered. The landing procedure for Mars is much smoother, as the atmosphere can be used for aerobreaking. Also for the power supply the conditions are more relaxed, as the mission can use solar panels and is not only dependant on battery power.

Other interesting targets for the microrover can be small bodies such as asteroids and comets. For these targets, of course, the conditions vary, e. g. the size of the bodies themselves, and thus the required design steps have to be carefully considered depending on the chosen body.

For missions with greater demand for mobility performances such as obstacle negotiation, rover range and greater payload capacity, the MRP Nanokhod rover provides an ideal design baseline. The technical challenges to miniaturise a practical flight implementation to 3.2 kg have been overcome and with a relaxation of mass and volume constraints the design allows for a safe and quick realisation for new mission scenarios. Possibilities of modification include replacing the tether system with internal power provision and RF communication, as well as upgrading the navigational capabilities. This offers a wide spectrum of missions which profit of an extended exploration range by implementing the Nanokhod.

Bibliography

[1] von Hoerner, H., Klinkner, S., *Explorationen unter den Aspekten der Rovertechnologie, Miniaturisierung und der Suche nach den Ursprüngen des Lebens*, DLR Explorationskonferenz, Dresden, Germany, Nov. 2006

[2] Bertrand R., *Conceptual Design and Flight Simulation of Space Stations*, Herbert Utz Verlag, München 1999

[3] Klinkner, S. et al., *Rover technologies developed for exploration of planetary surfaces*, DGLR International Symposium, Bremen, Germany, Sep. 2008

[4] Klinkner, S. et al., *Robotic exploration of planetary surfaces – Rover technologies developed for space exploration*, Research and Education in Robotics - EUROBOT 2008, Page(s): 193 – 206, Springer Verlag, Heidelberg, 2009 F

[5] Bertrand, R., Lamon, P., Michaud, S., Schiele, A., Siegwart, R., *The Solero rover for regional exploration of planetary surfaces*, Geophysical Research Abstracts, Vol. 5, 11850, European Geophysical Society, 2003

[6] Lee, C G–Y., Dalcolmo, J., Klinkner, S., Richter, L., Terrien, G., Krebs, A., Siegwart, R., Waugh, L., Draper, C., *Design and Manufacture of a full size Breadboard ExoMars Rover Chassis*, ASTRA 2006, Noordwijk, Nov. 2006

[7] Bernhardt, B., *Gipf Summary Report*, GIPF–TN–730–VHS–0, von Hoerner & Sulger GmbH, January 2006

[8] van Winnendael, M., Visentin, G., Bertrand, R., Rieder, R., *Nanokhod Microrover Heading Towards Mars*, ISAIRAS 1999, ESTEC, Noordwijk, the Netherlands

[9] Bertrand, R., *Micro–RoSA – Summary Report*, Issue 1, von Hoerner & Sulger GmbH, March 1999

[10] Bertrand, R., Dalcolmo, J., Klinkner, S., *RTPE– Final Report*, RTPE–54–10, Issue 1, von Hoerner & Sulger GmbH, June 2003

[11] Klinkner, S. et al., *Destination Moon and beyond for the Microrover Nanokhod*, DGLR International Symposium, Bremen, Germany, March 2007

[12] Lee, C.G.–Y., Klinkner, S., Hlawatsch, W., Wagner, C., Schreyer A.–M., Röser, H.–P., Schaefer, J., Schiele, A., Romstedt, J., *Mercury Nanokhod Rover – Hardware Realisation and Testing*, ASTRA 2006, Nordwijk, November 2006

[13] Schiele, A., Romstedt, J., Lee, C., Henkel, H., Klinkner, S., Bertrand, R., Rieder, R., Gellert, R., Klingelhöfer, G., Bernhardt, B., Michaelis, H., *The new NANOKHOD: Engineering model for extreme cold environments*, Robotics & Automation Magazine, IEEE, Volume 15, Issue 2, June 2008, Page(s): 96 – 107

[14] ESA, *BepiColombo – An Interdisciplinary Cornerstone Mission to the Planet Mercury*, System and Technology Study Report, ESA–SCI(2000)1, April 2000

[15] ESA/ESTEC, *BepiColombo Mission Requirements Document*, Issue 1, rev. 4, SCI–PF/BC/RS/01 (15.5.2001)

[16] Dunne, J., Burgess, E., *The Voyage of Mariner 10*, NASA, Jet Propulsion Laboratory, Washington D.C. 1978

[17] Davies, M., Dwornik, S., Gault, D., Strom, R., *Atlas of Mercury*, NASA, Prepared for the Office of Space Sciences, 1978

[18] Stramaccioni, D., *Mercury Environmental Specification (Part 1)*, Bepi-Colombo Definition Study, Reference SCI–PF/BC/TN/01, Issue 2.0, Revision 0, Date 4 July 2002

[19] Soerensen, J., Evans, H., *Mercury Environmental Specification (Part 2)*, BepiColombo Definition Study, Reference SCI–PF/BC/TN/02, Issue 2.0, Revision 0, Date 19. August 2002

[20] Heiken, G. et al., *Lunar Sourcebook*, Cambridge University Press, 1991

[21] ESTEC, *BepiColombo Payload Definition Document*, SCI–A/2002/007/Dc/CE, Issue 3, Rev. 1.1, 30 June 2003

[22] Bertrand, R., *RTPE Requirements Document*, RTPE–12–2.1, Issue 2.1, von Hoerner & Sulger GmbH, March 2003

[23] ESA, *Development of a compact Geochemistry Instrument Package Facility*, Statement of Work. SCI–ST/2002/002/SW/JR, 2001, Appendix 1 to AO/1–4175/02/NL/HB

[24] Bertrand, R., Lee, C., Klinkner, S., Henkel, H., *MRP Requirements Documents*, MRP–TN–110–VHS–3, von Hoerner & Sulger GmbH, March 2005

[25] Henkel, H., *GIPF Instrument Requirements*, GIPF–TN–130–VHS–1, Issue 1, von Hoerner & Sulger GmbH, October 2003

[26] Sjöholm, M., *RTPE Motor/Gearbox Assembly – Development Test Report*, Mecanex SA, Doc. RE–302.04–01, edition A, 4.10.02

[27] Schäfer, I., Klinkner, S., *Präzisionsgetriebe für die Raumfahrt*, Antriebstechnik 04/2006

[28] Harmonic Drive AG, *Konstruktionshandbuch*, 4. Auflage, September 2005

[29] Schäfer, I., *Technical Description and Performance Data MRP gear units*, Harmonic Drive AG, February 2005

[30] Rohloff, R.–R., Baumeister, H., Ebert, M., Münch, N., Naranjo, V., *Cryogenic actuators in ground–based astronomical instrumentation*, Proceeding contribution to Astronomical Telescopes and Instrumentation, 2004

[31] Lee, C., Klinkner, S., *Investigation of Motor Drive blocking: Mechanical*, MRP–TN–341–vHS, Issue 1.0, von Hoerner & Sulger, April 2006

[32] Anderson, M.J., Cropper, M., Roberts, E.W., *The Tribological Characteristics of Dicronite*, ESTL,ESR Technology Ltd. UK, 2007

[33] Klinkner, S. et al., *Experiences gained from the thermal vacuum tests of the Microrover Nanokhod*, ASTRA 2008, Noordwijk, Nov. 2008, *see paper in appendix G*

[34] Klinkner, S., *MRP Drivetest*, MRP–TN–430–VHS, Issue 0.1, von Hoerner & Sulger, January 2008

[35] Schlienz, U., *Der Schrittmotor der nicht ausrastet*, WEKA FACHMEDIEN GmbH, 2007

[36] Hughes, A., *Electric Motors and Drives*, Third Edition, Elsevier Ltd., 2006

[37] Arsape, *AM1020–ww–ee – datasheet*, www.arsape.com, Edition 23.07.2007

[38] Hillebrandt, S., Klinkner, S., *MRP – Coldtesting of brushless DC Motor*, MRP–TN–450–vHS, Issue 1.0, von Hoerner & Sulger, July 2008

[39] Wagner, C., Lee, C., *Cold Electronics Component Testing*, MRP–TN–233–vHS, Issue 1.1, von Hoerner & Sulger GmbH, March 2005

[40] Lee, C., *Power Transmission Comparison*, MRP–TN–231–1–vHS, Issue 1.0, von Hoerner & Sulger GmbH, January 2005

[41] Klingelhöfer, G. et al., *Jarosite and Hematite at Meridiani Planum from Opportunity's Mössbauer Spectrometer*, Science, Vol.306, No. 5702, 2004

[42] Morris, R. V. et al., *Mineralogy at Gusev Crater from the Mössbauer Spectrometer on the Spirit Rover*, Science, Vol. 305, No. 5685, 2004

[43] Wertheim, G., *Mössbauer Effect: Principles and Applications*, Academic Press, New York, 1964

[44] Bernhardt, B., Henkel, H., Krein, G., Wagner, C., *Study and Design of a Highly Integrated Communalised Payload System for a Planetary Exploration Rover*, HICOPS–TN–300, von Hoerner & Sulger, 15 August 2006

[45] Rohner, U. et al.,*Miniaturised TOF Mass Spectrometer, First Results*, EGS 2003, Nice, France, April 2003

[46] Weigand, B., von Wolfersdorf, J., *Wärmeübertragung – Manuskript zur Vorlesung*, Auflage, 2007

[47] Lienhard, J.H. IV, V, *A heat transfer textbook*, Phlogiston Press, third edition, version 1.24, Massachusetts, 2006

[48] Baehr, H. D., Stephan, K., *Wärme– und Stoffübertragung*, 4. Auflage, Springer Verlag, Heidelberg, 2003

[49] Kirschmann, R., *High Temperature Electronics*, IEEE Press 1999

[50] Dalcolmo, J., *RTPE Technological Constraints for Parts and Processes*, RTPE–13–1.0 Issue 1.0, von Hoerner & Sulger GmbH, June 2001

[51] Kirschman, R.K., *Die Attachment for $-120°C$ to $+20°C$ Thermal Cycling of Microelectronics for Future Mars Rovers: An Overview*

[52] Kirschman, R.K., *Extreme–Temperature Electronics Tutorial*, ©2002 – 2006 by Randall K. Kirschman, http://www.extremetemperatureelectronics.com

[53] Dalcolmo, J., *RTPE Verification Plan*, RTPE–24–3.0 Issue 3.0, von Hoerner & Sulger GmbH, September 2002

[54] Dalcolmo, J., *RTPE Verification Report*, RTPE–45–1.2 Issue 1.2, von Hoerner & Sulger GmbH, March 2003

[55] Evans, H. D. R., *Mercury Orbiter Radiation Dose*, Esa/estec/TOS–EMA/he/99–143, Issue 1, Rev. 0, ESTEC, TOS–EMA, 15 November 1999

[56] Gilmore, D., *Spacecraft Thermal Control Handbook, Volume I: Fundamental Technologies*, 2nd edition, The Aerospace Press, 2002

[57] Mastel Aluminium–Halbzeuge GmbH, *Werkstoffe nach DIN 1725/1*, http://www.mastel.de

[58] ECSS Secretariat, *ECSS–E–10–03A – Space engineering – Testing*, ESA Publications Division, Noordwijk, The Netherlands, 15 February 2002

[59] Schreyer, A.–M., Klinkner, S., Lee, C., Röser, H.–P., Schaefer, J., *Thermal analysis of the Nanokhod Microrover under Mercury environment test conditions*, IAC 2007, Hyderabad

[60] Klinkner, S., *Thermal vacuum test*, MRP–TN–440–TV–test, Issue 1.0, von Hoerner & Sulger GmbH, July 2008

[61] Barbé, J., *Mercury Thermal Model*, ESTEC–Noordwijk, (Aug. 2003)

[62] Klinkner, S. et al., *Lunar Exploration using Small Satellite with Micro Rover Technology*, 10th ISU Annual International Symposium, Strasbourg, France, 2005, *see paper in appendix H*

[63] Williams, D., *Mercury Fact Sheet*, NASA, NASA Goddard Space Flight Center, 30. November 2007

[64] Schaefer, J., Rudolph, S., *Satellite design by design grammars*, German Aerospace Congress 2003, Aerospace Science and Technology 9 (2005) 81–91, 25 September 2004

[65] Schöning, U., *Theoretische Informatik kurz gefasst*, BI Wissenschaftsverlag, Mannheim, 1993

[66] ALSTOM Power Technology Centre, *ESARAD user manual*, UM–ESARAD–024, Version 6.4, September 2004

[67] ALSTOM Power Technology Centre, *ESATAN User Manual*, UM–ESATAN–004, ESATAN 9.2, October 2004

A Constants and data

A.1 Physical and spaceflight constants

Speed of light in vacuum	c	299729458	m/s
Planck constant	h	$6.6260755 \cdot 10^{-34}$	J·s
Stefan–Boltzmann constant	σ	5.67051×10^{-8}	$W \cdot m^{-2} \cdot K^{-4}$
Astronomical Unit	AU	$1.4959787 \cdot 10^{11}$	m
Space sink temperature	T_{space}	3	K

A.2 Material data applied in the thermal model

The physical properties of the rover materials are given in table A.1.

Material	Density ρ [kg/m^3]	Specific heat capacity c_p [J/(kgK)]	Thermal conductivity λ [W/(mK)]	Comment
Aluminium	2770	875	130	
Steel	7900	500	4	for 20°C
PTFE	2240	960	0.23	
FR4	1200	1750	0.23	
PEEK	1550	1850	0.25	
Copper	8920	385	360	
Regolith	1300	800	0.01	

Table A.1: Material data

The optical properties for infrared and solar radiation are given in table A.2 and A.3.

Material	Infrared			
	Emissivity	Trans–missivity	Diffuse reflection	Specular reflection
Aluminium	0.65	0.0	0.35	0.0
Aluminium – Alodine 1200	0.12	0.0	0.88	0.0
Stainless steel	0.3	0.0	0.7	0.0
PTFE	0.89	0.0	0.11	0.0
FR4	0.89	0.0	0.11	0.0
PEEK	0.89	0.0	0.11	0.0
Copper	0.76	0.0	0.24	0.0
Regolith	0.9	0.0	0.1	0.0

Table A.2: Surface properties with infrared radiation

Material	Solar			
	Absorption	Trans–missivity	Diffuse reflection	Specular reflection
Aluminium	0.5	0.0	0.5	0.0
Aluminium – Alodine 1200	0.5	0.0	0.5	0.0
Stainless steel	0.47	0.0	0.53	0.0
PTFE	0.75	0.0	0.25	0.0
FR4	0.72	0.0	0.28	0.0
PEEK	0.75	0.0	0.25	0.0
Copper	0.70	0.0	0.30	0.0
Regolith	0.9	0.0	0.1	0.0

Table A.3: Surface properties with solar radiation

A.3 Physical and chemical aspects of the planet Mercury

The Mercury properties [63], which are of interest for this thesis are the following:

	Mercury
Mass	0.3302×10^{24} kg
Volume	6.083×10^{10} km^3
Equatorial radius	2439.7 km
Polar radius	2439.7 km
Volumetric mean radius	2439.7 km
Mean density	5427 kg/m^3
Surface acceleration	3.70 m/s^2
Escape velocity	4.3 km/s
Bond albedo	0.119
Solar irradiance	9126.6 W/m^2
Black–body temperature	442.5 K
Semimajor axis	57.91×10^6 km
Sidereal orbit period	87.969 days
Perihelion	46.00×10^6 km
Aphelion	69.82×10^6 km
Mean orbital velocity	47.87 km/s
Maximum orbital velocity	58.98 km/s
Minimum orbital velocity	38.86 km/s
Orbit inclination	$7.00°$
Orbit eccentricity	0.2056
Sidereal rotation period	1407.6 hrs
Length of day	4222.6 hrs
Surface pressure	10^{-15} bar

B Trade–off of gear combinations

Table B.1 and B.2 show possible gear-motor combinations which were traded against each other for the MRP project.

The gear configuration consisting of the two Harmonic Drive units (cases 1.1–1.3), are preferable for the reduction of the backlash, having two gear units with zero backlash and reducing the small backlash of the crown gear by the ratios 80:1 and 160:1. Nevertheless the input torque of the Micro Harmonic Drive is fairly high for the given values, based on the Drive specifictions in case of industrial use. Furthermore the resulting torques of these gear configurations is only sufficient for the application on Mercury, having a lower gravity. For the testing of the rover model an application on Earth is essential, this is why these configurations are discarded.

In the cases 2.1–2.3 the gear configurations are consisting of a planetary gearhead a crown gear and a harmonic drive unit. As the input speed value for the planetary gearheads of the two stepper motors is quite low, the output speeds for those are limited to 3.06 m/h (for the AM1020) and to 1.96 m/h (for the AM0820). These speed values are is both lower than the 5 m/h defined by the requirement document. However the input speed requirement of the gearhead bases on the industrial qualification for the gear-units, assuming a much higher lifetime as well as a longer use. The planetary gearhead for the EC6 motor does not have such a low limitation for the input speed, and thus the output speed for this configuration even slightly exceeds the required speed of 5 m/h The output torque values are higher than for the cases 1.1–1.3., however only case 2.1 produces sufficient torque for testing also the more demanding operations of the rover under Earth gravity conditions. The torque values for case 2.2 and 2.3 are still rather on the limit for what is required for Earth conditions.

In order to meet the speed requirement of MRP, it can be considered to qualify the planetary gearhead for the required speed value in a next step towards the flight model, basing on the limited mission lifetime. Another option for meeting the requirement is to use the gear assembly of case 2.3 with the EC6 motor, which gives high enough torque for the environment on Mercury, however only insufficient torque for the testing of the demanding manoeuvres on Earth.

Based on these considerations, the motor-gear-assembly of case 2.1 has been

Case 1.1	AM1020 motor exit	Crown gear exit	MHD 8–160 exit	HDUC 5–80 exit	Track wheel/ Track exit
Speed	20000 rpm	6666.7 rpm	41.7 rpm	0.5 rpm	4.9 m/h= 1.4 mm/s
Translation ratio	–	3:1	160:1	80:1	
Efficiency	0.69	0.85	0.5	0.5	
Min. torque [mNm]	0.2	0.51	40.8	1632	
Case 1.2	AM0820 motor exit	Crown gear exit	MHD 8–160 exit	HDUC 5–80 exit	Track wheel/ Track exit
Speed	16000 rpm	5333.33 rpm	33.3 rpm	0.4 rpm	3.9 m/h= 1.1 mm/s
Translation ratio	–	3:1	160:1	80:1	
Efficiency	0.69	0.85	0.5	0.5	
Min. torque [mNm]	0.2	0.51	40.8	1632	
Case 1.3	EC6 motor exit	Crown gear exit	MHD 8–160 exit	HDUC 5–80 exit	Track wheel/ Track exit
Speed	55000 rpm	18333.3 rpm	114.6 rpm	1.4 rpm	13.5 m/h= 3.8 mm/s
Translation ratio	–	3:1	160:1	80:1	
Efficiency	0.69	0.85	0.5	0.5	
Min. torque [mNm]	0.2	0.51	40.8	1632	

Table B.1: Possible drive combination including a micro Harmonic Drive

chosen for the MRP rover (Table 4.2). With this new motor-gear-assembly the backlash was reduced with respect to the RTPE-configuration and thus the accuracy of the instrument positioning could be increased. By using this configuration, the torque requirements for the rover manoeuvres are fulfilled also under Earth gravity conditions. In order to fully meet the speed requirements

Case 2.1	AM1020 motor exit	Planetary gear exit	Crown gear exit	HFUC 5–100 exit	Track wheel/ Track exit
Speed	5000 rpm	78.13 rpm	32.17 rpm	0.32 rpm	3.03 m/h= 0.84 mm/s
Translation ratio	–	64:1	34:14	100:1	
Efficiency	0.69	0.7	0.85	0.5	
Min. torque [mNm]	0.7	31.36	64.7	3236.8	
Case 2.2	AM0820 motor exit	Planetary gear exit	Crown gear exit	HDUC 5–50 exit	Track wheel/ Track exit
Speed	8000 rpm	31.25 rpm	10.42 rpm	0.208 rpm	1.96 m/h= 0.54 mm/s
Translation ratio	–	256:1	3:1	50:1	
Efficiency	0.69	0.69	0.85	0.5	
Min. torque [mNm]	0.17	30.3	76.57	1914.33	
Case 2.3	EC6 motor exit	Planetary gear exit	Crown gear exit	HDUC 5–50 exit	Track wheel/ Track exit
Speed	18000 rpm	81.45 rpm	27.15 rpm	0.54 rpm	5.11 m/h= 1.42 mm/s
Translation ratio	–	221:1	3:1	50:1	
Efficiency	0.69	0.6	0.85	0.5	
Min. torque [mNm]	0.23	30.5	77.77	1944.25	

Table B.2: Possible drive combinations including planetary gearhead

there are the two possibilities, to either augment the speed (and having sufficient torque for Mercury conditions), by using the EC6 with gearhead, or to qualify the planetary gearhead of the AM1020 motor for the higher speed for the limited mission duration.

C Thermal model software

The software packages that were used to build the thermal mathematical model of the Nanokhod rover, as described in section 6.2.3, are illustrated to more details in the following paragraphs.

C.1 Design compiler 43

The design compiler acts as a front end to computation and visualisation tools [64]. It is used to translate the necessary data and the internal interfaces of the system that is to be analysed into a network and analysing it by using external analysis tools.

For the design process with the compiler 43 first a *design grammar* has to be defined. This design grammar originates from the science of computer languages [65], like C++. In this context there are defined items of vocabulary (see below) which are combined through a set of rules (see below) to the description of the problem, which then would be the *sentence*. In contrast to a language grammar, the design grammar is able to combine the vocabulary not only to a *linear sentence*, but also to more complex network structures, which are necessary when considering engineering solutions of a system.

With the design grammars summarised to a design programme (see below), the compiler will construct a design graph, which represents the network problem to be solved by the analysis tool. The analysis can then be carried out by activating the necessary application specific plug–in functions. These plug–in functions prepare the necessary output files, which are then used by the external analysis tools.

Thus, it is possible to have a network type problem, which may be solved for several different kinds of analysis. This of course means the vocabulary and the rules have to be prepared for the specific analysis that is carried out by defining the necessary input data. In the framework of this thesis the Nanokhod system network was only prepared for the thermal analysis, but the analysis could be easily expanded to e. g. structural analysis, thermal stress analyses, etc..

Vocabulary The vocabulary of a design grammar represents a design component and contains all functional physical and geometrical properties necessary for the analysis of this component. The items of vocabulary are the nodes of the network and can contain parameters, variables and operations that are describing any dependency of the component.

The parameters can either be defined within the vocabulary or by rules during the analysis of the system. The interfaces between items of vocabulary can exchange parameters and variables through the defined ports (interface specifications) between the components. The connection between two components is only possible if the port specification is the same for these two parts.

Grammar The rules of a design grammar define the connections between the items of vocabulary. The definition of a rule consists of two parts which is the conditional part and the generative part. The conditional part has to be given by the system, so the generative part is then carried out.

Depending on the definition of the rules as well as on the order of there execution, such rules can be used to set up a network or to reiterate an existing network. Furthermore the rules can be defined in a general way so they can be used for the grammar of all types of system.

Programme The programme will add all information defined in the vocabulary and the grammar together by creating a design graph which represents the whole system network. The programme is always initiated by an axiom, which is setting the first condition on which the following rules can base. This axiom is basically representing the environment or the requirements which have been defined for the system.

C.2 ESARAD

The ESARAD software is a thermal radiative analysis tool carrying out the pre– and post–processing for the ESATAN thermal analysis package [66].

It uses either Monte–Carlo ray tracing (MCRT) or the matrix methods (MM) for the analysis of the radiative heat exchange from surface to surface. The advantages of MCRT are the high accuracy and no special requirements have to be fulfilled for the valid use of it. It thus also allows the calculation of specular reflection. The MM can only be applied for isothermal surfaces with constant thermo–optical properties and is only valid for diffuse reflection. However, the MM only performs matrix operations, which has the advantage of a much faster evaluation of the radiative couplings. For the evaluation of the radiative couplings within the model described in this thesis the MCRT was used because of its higher accuracy.

ESARAD provides the ability to set up the optical geometry of the model that shall be analysed. This is done by combining various primitive shapes to represent the physical components and assemble them to a full model geometry. For the model considered in this thesis, the geometry was build with the help of the compiler language and read into ESARAD in order to calculate the radiative heat exchange parameters and prepare the model for ESATAN.

For the thermal analysis the geometry of the rover is represented by a simplified geometrical model which is then decomposed into discrete isothermal nodes. The model input includes the geometry of the model for radiative couplings, defined conductive couplings, defined heat fluxes and, if valid, orbit data. The in–orbit calculations base on the definition of spacecraft trajectories, attitudes and pointing.

The thermal model assigns the following thermal properties within the nodes: Thermo–optical properties, which are emissivity, absorption, transmissivity and diffuse and specular reflectivity, for infrared as well as solar radiance; and bulk properties, which are density, heat capacity and thermal conductivity.

The programme supports the analysis of radiation from different surface types such as diffuse, specular and transmissive radiative heat transfer.

Within ESARAD the heat transfer data of the model – namely heat flux, conductance to adjacent nodes and radiative exchange factors – are calculated. These values are then used as input data for ESATAN.

C.3 ESATAN

The ESATAN software is used to calculate temperature distributions of the model for single components or whole systems by analysing the thermal network [67]. The model to be analysed is specified as a quantity of nodes which are linked within a thermal network. The network contains information about the node properties as well as the properties of the node linkages. This network is defined and calculated in ESARAD and transferred into a system of differential equations – the so called thermal mathematical model. The coefficients of the equations depend on the physical properties of the model. The thermal mathematical model (TMM) is then formatted into an input file for ESATAN based on the selected model solution.

Using that TMM ESATAN enables the calculation of the steady–state or transient temperature distributions. The differential equations are solved using a finite differences algorithm.

Prior to the solving the set of differential equations the input file is going through a pre–processor. It checks the syntax and organises the data and operation blocks for the later calculation.

ESATAN has the facility to handle not only radiation and convection as was needed for the model in description, but also convectional heat transfer. Also possible but not applied to the Nanokhod analysis are phase–change phenomena such as boiling, condensation, melting or solidification.

D Development schedule

The suggested schedule for a flight model development of the Mercury design of the Nanokhod rover, which is directly following the MRP project and the additional activities of this thesis is shown in the graphs D.1 and D.2.

The main development steps are as follows:

- **Environmental tests:** For a good baseline of the following development steps the rover model on hand shall be used for extensive environmental tests, in order to have a good understanding of the rover system and any related issues.
- **EQM – phase:** The EQM design is divided into the following sub–tasks:
  - **Design:** Based on the experiences gained from the current design and the environmental tests the design of the EQM model can be conducted concentrating on the drive unit and on the accommodation of the scientific geochemical payload
  - **ASIC development**
  - **Desktop model:** A desktop model shall externally test the key components of the rover, which are the drive unit, the accommodated payload and the electronics.
  - **Software** Development of the GSE and control Software
  - **Manufacturing** of the mechanical and electronic components, including the integration of the rover EQM model.
  - **Testing:** Carrying out all necessary environmental tests with qualification levels.
- **Flightmodel:** The flight model activity is subdivided into the following tasks
  - **Design update:** Including the integration of all flight components into the design.
  - **Software** Development of the flight software for the GSE and control.
  - **Manufacturing** of the flight model electronics and mechanical components
  - **Testing:** Environmental tests of the flight model to acceptance levels
- **Flight spare activity:** The flight spare activity consists of the same development steps than the flight model without the design updates, as the flight spare will base on the exactly same design as the flight model.

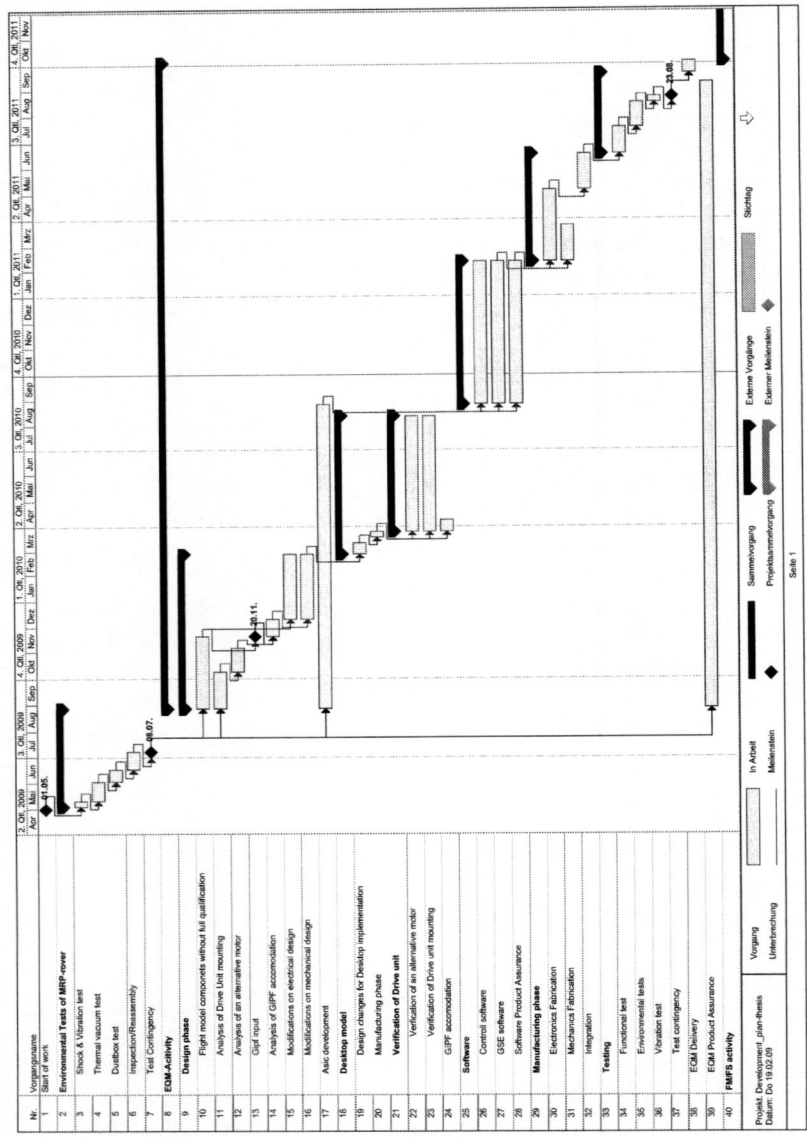

Figure D.1: Development schedule for a Nanokhod flight model – environmental test– and EQM–activity

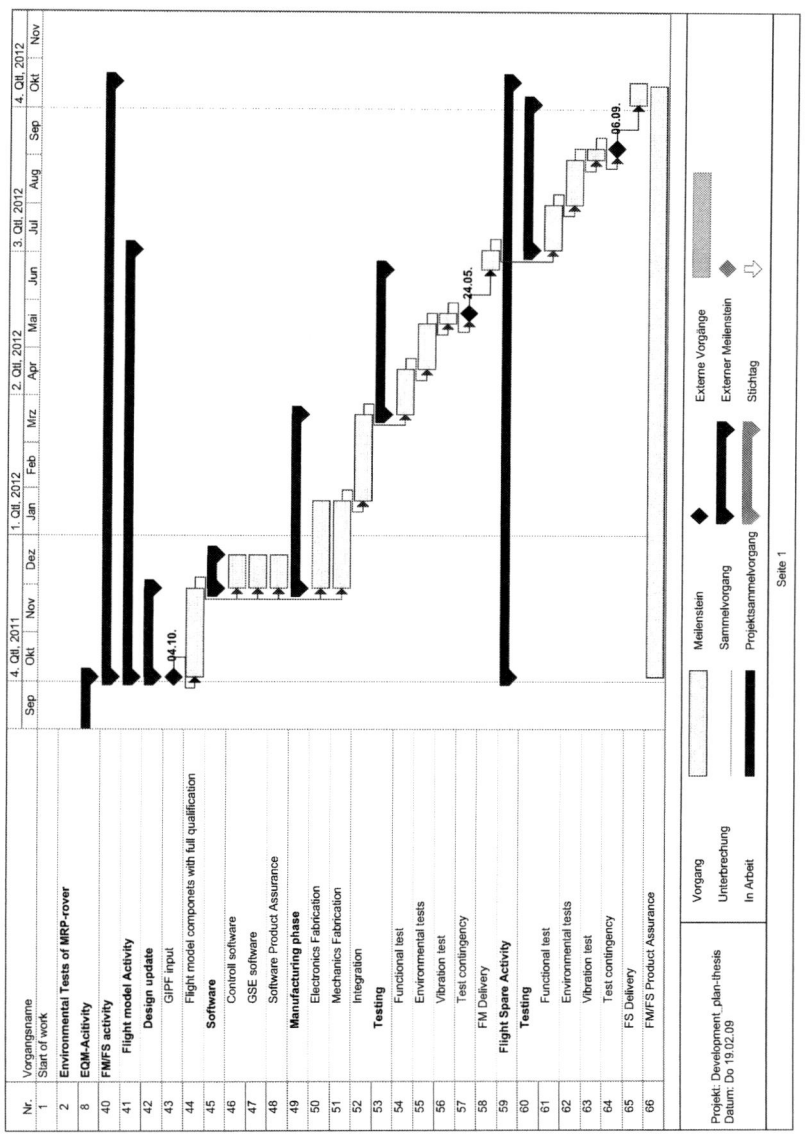

Figure D.2: Development schedule for a Nanokhod flight model – flight and flight spare model

D Development schedule

E List of publications

Publications

- S. Klinkner, C. Lee, H.–P. Röser, G. Klingelhöfer, B. Bernhardt, I.Fleischer, D. Rodionov, and M. Blumers *Robotic exploration of planetary surfaces – Rover technologies developed for space exploration*, Research and Education in Robotics - EUROBOT 2008, Page(s): 193 – 206, Springer Verlag, Heidelberg, 2009
- André Schiele, Jens Romstedt, Chris Lee, Hartmut Henkel, Sabine Klinkner, Reinhold Bertrand, Rudi Rieder, Ralf Gellert, Göstar Klingelhöfer, Bodo Bernhardt, Harald Michaelis, *The New Nanokhod Exploration Rover for Extreme Planetary Environments*, Robotics & Automation Magazine, IEEE, Volume 15, Issue 2, June 2008, Page(s): 96 – 107

Conferences

- Jessberger, E.K., I. Rauschenbach, H. Henkel, S. Klinkner, H.-W. Huebers, S.G. Pavlov, *GENTNER – a miniaturized laser instrument for planetary in-situ analysis*, International Workshop Europa Lander: Science goals and experiments, 9.–13. February 2009
- Sabine Klinkner, Chris Lee, Carsten Wagner, Michael Lengowski, Hans–Peter Röser, Pascale Bourlier, *Experiences gained from the thermal vacuum tests of the Microrover Nanokhod*, ASTRA 2008, Noordwijk, Nov. 2008
- Klinkner, S., Lee, C.G.–Y., Wagner, C. Hlawatsch, W., Dalcolmo, J., Röser, H.–P., *Rover technologies developed for exploration of planetary surfaces*, DGLR International Symposium, Bremen, Germany, Sep. 2008
- S. Klinkner, C. Lee, H.–P. Röser, G. Klingelhöfer, B. Bernhardt, I. Fleischer, D. Rodionov, and M. Blumers *Robotic exploration of planetary surfaces – Rover technologies developed for space exploration* EUROBOT 2008, 21.–24.5.2008, Heidelberg *see paper in appendix F*

- Schreyer, A.–M., Klinkner, S., Lee, C., Röser, H.–P., Schaefer, J. *Thermal analysis of the Nanokhod Microrover under Mercury environment test conditions*, IAC 2007, Hyderabad
- Klinkner, S., Lee, C.G.–Y., Wagner, C. Hlawatsch, W., Schreyer A.–M., Röser, H.–P., *Destination Moon and beyond for the Microrover Nanokhod*, DGLR International Symposium, Bremen, Germany, Mar. 2007
- H. von Hoerner, S. Klinkner, *Explorationen unter den Aspekten der Rovertechnologie, Miniaturisierung und der Suche nach den Ursprüngen des Lebens*, DLR Explorationskonferenz, Dresden, Germany, Nov. 2006
- Lee, C G–Y, Dalcolmo, J, Klinkner, S, Richter, L, Terrien, G, Krebs, A, Siegwart, R, Waugh, L, Draper, C, *Design and Manufacture of a full size Breadboard ExoMars Rover Chassis*, ASTRA 2006, Noordwijk, Nov. 2006
- Lee, C.G.–Y., Klinkner, S., Hlawatsch, W., Wagner, C., Schreyer A.–M., Röser, H.–P., Schaefer, J., Schiele, A., Romstedt, J. *Mercury Nanokhod Rover – Hardware Realisation and Testing*, ASTRA 2006, Noordwijk, Nov. 2006
- A. Schiele, J. Romstedt, C. Lee, H. Henkel, S. Klinkner, R. Rieder, R. Gellert, G. Klingelhöfer, B. Bernhardt, H. Michaelis, *The new NANOKHOD: Engineering model for extreme cold environments*, 8th International symposium on Artificial Intelligence, Robotics and Automation in Space, München, Germany, September 2005
- S. Klinkner, R. Laufer, T. Graf, M. Nagy, H.–P. Roeser, *Lunar Exploration using Small Satellite with Micro Rover Technology*, 10th ISU Annual International Symposium, November 30 – December 2, 2005, Strasbourg, France, 2005

F Eurobot Conference Paper 2008

Robotic exploration of planetary surfaces – Rover technologies developed for space exploration

Authors: S. Klinkner[1,2], C. Lee[1], H.–P. Röser[2], G. Klingelhöfer[3], B. Bernhardt[1], I. Fleischer[3,4], M. Blumers[3]

EUROBOT Conference 2008, Heidelberg 21. – 24. May 2008

[1] von Hoerner & Sulger GmbH, Schlossplatz 8, 68723 Schwetzingen, Germany, Email: klinkner@vh–s.de
[2] Universität Stuttgart, Institut für Raumfahrtsysteme, Stuttgart, Germany
[3] Johannes Gutenberg Universität Mainz, Institut für Anorganische und Analytische Chemie, Mainz, Germany
[4] Space Research Institute IKI, Moscow, Russia

F.1 Abstract

Mobility is a key feature for any science mission and for space exploration in general. Missions with mobile systems provide a much wider spectrum of outcomes by employing a higher number of samples within an increased area of exploration. The additional degree of freedom of a Rover in comparison to a lander or even a robotic arm allows the mission to be flexibly adapted to the landing site as it is encountered.

Nevertheless, Rover vehicles developed for the exploration of planetary surfaces are extreme complex systems, which have to be specialised for the environmental conditions they are dedicated for. With the variation of the environmental conditions on missions to different target planets, the requirements are varying for the landing system, the rover as well as the payload.

Since 1989 the company *von Hoerner & Sulger* is doing research in the field of robotic systems and planetary exploration. Given that, the company is in the mean time well situated in the development and manufacture of rover and established a good cooperation with academic institutes. The company gained the experience to develop the matching rover chassis for a variety of mission scenarios:

The Nanokhod Rover is a small mobile scientific platform, designed to transport a package of scientific instruments and to carry out in–situ measurements of rocks and small craters in the vicinity of the landing point. The Microrover has a volume of $160 \times 65 \times 250$ mm, it weighs 3.2 kg including a payload mass of 1 kg and has a peak power need of max. 5 W. The Nanokhod is a tethered system that uses the Lander for power supply and as a data relay to Earth. The Nanokhod has recently been designed to withstand the demanding requirements of a flight model on a mission to Mercury. Based on this design, an engineering–level hardware model was built which is suitable for environmental testing, preparing the Rover design for a variety of possible future missions.

The Solero Rover is an innovative Minirover concept, designed for regional exploration of a planetary surface. The vehicle has a passive chassis concept with exceptional climbing abilities, which provides the ability to adjust to all kinds of terrains and thus minimises control needs.

The company vH&S is leading already the second ExoMars Rover chassis breadboard design and manufacturing activity, which is part of the rover development for the first European Mars rover. The second breadboard is designed and built by vH&S GmbH in collaboration with DLR and two Swiss collaborators. The ExoMars Rover for the ESA Cornerstone Mission Aurora, will be a mobile Laboratory having an Exo–biology Payload (Pasteur), including a geochemical package, and carrying a drill that is reaching probes up to a depth of two meter.

This paper describes the gained experiences and most important aspects of a rover design for the purpose of planetary exploration. In addition it presents the newest designs and the manufactured models in relation to their missions ...

F.2 Introduction

von Hoerner & Sulger GmbH (vH&S) is a small scale enterprise (SME) that is working since 1971 successfully in the field of space exploration. This covers scientific space instruments, rocketborne systems, cameras and sensors for space applications as well as robotic systems for planetary exploration. The areas of operation of vH&S include concept finding, feasibility studies, development, fabrication, and qualification of space systems for applications in extraterrestrial missions.

Since its foundation the company has produced more than 10 flight–qualified scientific space instruments. Famous examples from the past are the first–ever mass spectrometers for cometary dust, PIA and PUMA (1 & 2), which were built between 1981–1984 for ESA's Giotto mission and for the Russian Vega 1 & 2 missions. These three experiments were the first and only ones that met comet Halley in 1986, giving insight into its chemical composition. After these brilliant results, vH&S became the prime company to provide mass spectrometers for both NASA and ESA missions.

This is only one example how vH&S has earned profound experience in the space mission activities and a worldwide high profile reputation in the space business, both in prime and subcontractor roles.

Based on this experience in the sector of space missions the company has expanded its portfolio in the late 1980's with the development of systems for in–situ planetary exploration. Since vH&S is active in this field it was involved in the development of various rover chassis concept matching for distinct mission scenarios. These chassis concepts cover the rover categories from a highly–integrated Mircorover to a big rover with a mass of up to 220 kg.

The missions scenarios vary from very hot to very cold environmental conditions, for targets with and without atmosphere and for planetary surfaces which are rough and rocky or smooth and covered with fine dust particles called regolith, namely environments on the planets Mars and Mercury, or our Moon. The most prominent example that made vH&S a leading European name for robotics is the *Nanokhod* Microrover, with a mass of only 3.2 kg including 1 kg of scientific payload mass. On the other end of the rover size scale vH&S is involved in ExoMars activity that is building the rover for the European Mars mission being part of the ESA Aurora programme.

F.3 Mobility for space exploration missions

Current trends in space exploration aim for mobile systems, see also [1]. In order to fulfil the whole range of scientific interests and to have a well–funded scientific examination of a planetary surface more than one sample has to be considered. This requires a mobile system to reach the different areas of interest.

A mobile system means that either several samples of different materials have to be collected and transported to the instrument or the instrument itself has to be moved to the different sample sites. A wide spectrum of results gives a wider view of the explored target; either way though requires the ability of movement. The application of a rover which transports instruments and carries out in–situ analysis has the advantage that the range of the rover system is farther.

The mobility aspect of the rover makes it possible to explore with a single space system a whole region of up to some tens of km around a landing point depending on the implemented system. This regional exploration provides a deeper knowledge and allows a more general characterisation of certain areas of a planetary surface. The mobility of such a system gives the scientific explorers the freedom to reach and examine specific points of interest. This offers also the opportunity to carry out a systematic exploration of the region. Depending on the system used and on the environmental conditions encountered this can even compensate in some cases the uncertainty of the landing ellipse.

For the mobility in space infrastructure the following development can be observed. For nearby planetary bodies which have been explored before, the implemented mobile systems are developed with an increasing reach (m $>$ km). While for bodies which require a long transfer and which are examined for the first time, the focus is rather on a system with very low mass. Mission success in this case is to gain in–situ analysis in the near surroundings of the landing point.

A variety of rover systems have been developed in the previous years in order to accomplish all needs for the exploration of target surfaces. Theses systems differ in their range, complexity, their demands on control, power needs etc. Thus a rover system can be adjusted to the expected mission scenario.

F.4 Rover technology for science missions

There are several performance drivers for rover used for space applications. One of the main drivers is the perception capability. This means either their ability to recognise and understand the environment they move through, or the identification of the targets they take samples of. In other words the rover has to percept the near surroundings, by seeing it, processing the information they have seen, and than analyse what this means for the further action. This routine has to be done for each segment of movement as well as for the taking or measuring of samples.

A further driver closely related to this are of course the mobility capabilities. This is necessary in order to reach all scientific targets, when moving through all kinds of terrain the rover might encounter. But the mobility aspects become also more important for the processing and handling of samples to prepare them for analysis. A rover will have to enable a close contact between an instrument and the sample. It will have to grasp and collect samples; it might even have to dig, grind or drill, to prepare the sample for the scientific analysis with certain instruments.

Especially for long distance mission targets with long signal delays, a further driver for space rover are their operational capabilities. For rover missions, there is a high demand for autonomy, in order to cope with hard real–time situations, as well as to not put any additional constraint on the mission duration because of signal delays. Furthermore the need for a high level of ground interaction is also always a cost factor for the mission. This demand for autonomy includes also autonomous decision making in an unknown and unstructured environment. The rover system has to be able to cope with unexpected situations and has to provide solutions for any contingency. Autonomous operation implemented in space rover also simplifies the payload (P/L) operations as well as the scientific target selection.

A rover system equipped with a high number of sensors for scientific analysis can utilise these capabilities also for the planning of single operations. Like that the rover can use instruments facilities for mission planning aspects. This use of synergy effects enhances the system efficiency, by reducing any additional instrumentation and thus system overhead by a careful integration of the instrumentation into the system.

Autonomy makes the rover adaptable to a variety of situations and missions and provides the necessary flexibility which the systems needs in a completely unknown environment. A space exploration approach in the rover system development can be achieved, if the rover is even developed together with the scientific payload, in order to optimise synergies.

F.5 Microrover Nanokhod for the Mercury surface exploration

Since 1992 the company *von Hoerner & Sulger GmbH* has the leading role in the development history of the Nanokhod Rover. The Rover is based on an originally Russian design, but has since undergone a significant development to provide a practical flight implementation, for example the accommodation of scientific payload instruments and the building of prototypes to test the mobility performance, like step and slope climbing abilities and sealing concepts to prevent the Rover inside from regolith.

The latest development is now a Nanokhod design which is able to cope with the challenging requirements for a flight model on the Mercury surface. A hardware model of this design is realised for environmental tests such as vibration, shock and thermal vacuum to the extreme requirements of a Mercury mission profile. The project to develop a mature Rover for the Mercury environment is based on the ESA Cornerstone mission BepiColombo.

Initially, the BepiColombo mission was still including a Surface element, and the Nanokhod Rover was selected to be part of it as a mobile scientific platform – the Mercury Robotic Payload (MRP). Unfortunately the Surface element, was cancelled from the BepiColombo mission and with it the flight opportunity of the Rover. Nevertheless, it was decided to continue with the development to prepare the Rover for future missions by solving detailed technical problems of a real implementation and thus gaining a better understanding. Also because the technology developed for the challenging nature of the Mercury environment is considered to be applicable with moderate modifications on a variety of other planetary bodies, with and without atmosphere, like Mars or Moon.

Small systems are required due to a limitation of financial resources and energetically more demanding missions. As a small and mobile system the Nanokhod thus fulfils two trends – mobility and low mass – for future missions, with additionally a very good payload mass to system mass ratio. And now with the design meeting the demanding flight model requirements for the conditions of a Mercury mission the Rover is very well prepared for a variety of future mission scenarios.

F.5.1 Mercury mission profile

After some initial studies of the Rover mission under day– as well as night side conditions of Mercury, the landing site was decided to be on the night side of the planet, where the absence of the sun's radiations reduces the thermal range and makes the mission feasible. Despite this the environment remains severe, coupling a high vacuum with surface temperatures estimated to $-180°$ C. However the night side landing site causes the disadvantage that all energy must be supplied chemically.

The MRP Nanokhod shall be able to move across the Mercury surface which is expected to consist of fine regolith similar to the Moon surface with a speed of 5 metres/hour and to negotiate steps of 10 cm height and trenches of 10cm width. The mission foresees one in–situ analysis per Earth day with each of its three instruments: Microrover camera (MIROCAM), APXS and Mössbauer Spectrometers. The Rover shall operate near–autonomously due to a single communication period once a day.

F.5.2 Design drivers resulting from the Mercury mission scenario

This mission profile induces as main design drivers mass and volume for the whole landing system. With an interplanetary journey between Earth and Mercury that requires a high amount of fuel to enter the Mercury orbit, and with a landing scenario which has to rely only on chemical propulsion, it becomes obvious that every kilogram to land on the planet is very cost intensive. This is why the mass and the volume of the Lander and thus the Rover have to be reduced to a minimum in order to make the mission feasible. Opposing this, the Rover has to be designed to be robust enough to cope with the vibration and shock environment of such a landing scenario. In order to reduce the amount of chemical propulsion to a minimum the landing shock is expected to be 200 g for a duration of up to 20 ms.

Also related to the mass limitation issue is the energy consumption of the Rover. The power is provided from the Lander's batteries and with a given mission duration, the consumed energy relates directly to the battery capacity and thus to the battery mass. This makes the energy consumption to be the next significant design driver of the Rover system.

During the mission, the Rover has levels of different energy consumption, depending on the rover activity. Based on an analysis of the mission and the duration of the different activities the energy consumption of the rover has to be optimised, in order to decrease the overall power needs as far as possible. In the case of the Nanokhod for the Mercury mission the energy consumption has to be as low as possible for the measurements phases as well as for the non–active periods of the Rover.

A low power consumption of the system can also be realised by implementing a passive thermal control and mechanisms which do not require power to maintain their state. Passive thermal control in this instance means that the Rover does not attempt to maintain its temperature to a set level but the Rover is allowed to heat up and cool as defined by the mechanical design and selected materials and finishes. As there may be extended periods when the Rover is powered off, all components should be able to function from the surface temperature upwards.

F.5.3 Overview of the Nanokhod design

A brief overview of design is given in the following paragraph. For a more detailed description of the Rover design including its instruments please refer to [2]. The main components which are generic to all Nanokhod Rovers are:

- Two locomotion units (LU) enclosed by walls and the driven caterpillar tracks which provide the method of locomotion
- The tether unit (TU) which rigidly attaches both locomotion units and contains the spools from which the tether wire is deployed
- The payload cabin (PLC) containing the scientific payload instruments
- Arms connecting the PLC to each of the LU giving the PLC two degrees of freedom allowing the instruments to place next to sample sites and for the PLC to act as an extra limb for negotiating obstacles.
- Four internal drive units used to drive the caterpillar tracks and position the arms relative to the LU and the PLC.

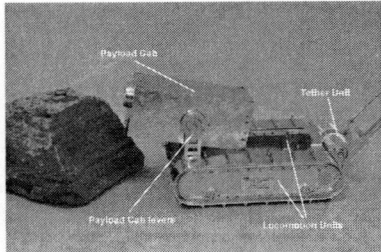

Figure F.1: Main components of the Nanokhod rover

The main components are identified on the current design in Figure F.1. The

Table F.1: The main dimensions of the Nanokhod rover

Nanokhod properties	
Mobility:	
Overcome obstacles up to:	0.1 m
Locomotion speed:	2.7 m/h
Electrical Power:	
During movement:	5.7 W peak
Other modes:	1.3–3.4 W
Rover Dimensions	
Mass including P/L and module on Lander:	< 3.2 kg
Size:	$< 250 \times 160 \times 65$ mm

overall structure has been upgraded to withstand the rigours of vibration and shock with the inclusion of four rigid yokes in the LU's. Analysis has been performed on all components to ensure that they are compatible to the mechanical and thermal environment. A completely new drive system based on a similar concept for all four drives within the Rover has been implemented, which was developed in a close cooperation with the Harmonic Drive AG. Due to the high vacuum environment it is not possible to use standard DC motors for extended durations and so the Faulhaber AM1020 stepper motor was selected as the motor for the drives. Dicronite dry lubricant was used on both the Harmonic Drive and crown gear the application of which had been tested at Harmonic Drive AG. The other components were supplied pre–lubricated with MOS2 by the manufacturer.

Electronically the system is has been partitioned into a number of nodes each of which perform a distinct function. Power for the drive system nodes is supplied by a 28V line which is also controlled by the tether interface node. When the power is removed from this line all drive unit nodes are powered off, minimising the power consumption during instrument operations.

F.5.4 Conclusions for the Microrover

In Table F.1 the main parameters of the highly integrated Nanokhod are listed. Although a mission to the Mercury surface is currently unlikely, the Moon has now become very popular in consideration for proposed visions, but also other targets stay possible. The night side environmental conditions of the Moon are similar to Mercury and would allow easy adaptation of the current concept. For a dayside landing the new Nanokhod model is still very applicable although new attention would need to be given to the thermal design, having the thermal model already on hand. For missions with greater demand for mobility performance such as obstacle negotiation, Rover range and greater payload capacity, the MRP Nanokhod Rover provides an ideal design baseline.

The technical challenges to miniaturise a practical flight implementation to 3.2 kg have been overcome and with a relaxation of volume and mass constraints the design allows a safe and quick realisation of the new mission scenarios. Possibilities of modification include replacing the tether system with internal power provision and RF communication as well as upgrading the navigational capabilities.

The MRP Nanokhod Rover is a huge advance towards a practical flight model of a highly integrated miniaturised mobile payload for planetary exploration despite limited resources that were available. Solutions for open issues have been implemented allowing the Rover to be subjected to the currently ongoing thermal vacuum testing and further environmental tests.

F.6 Minirover Concept Solero for regional exploration

The mobility requirements for a chassis of the Minirover type call for a typical travel distances of several kilometres with an average locomotion speed of up to 100 m/h (EXOMARS). It is evident that navigation and control of such a vehicle needs another approach as compared to local mobility systems that remain in the vicinity of a stationary lander. Autonomous navigation is required to enable at least the typical operational increment between two consecutive ground control interventions, which is one sol. Besides the principal autonomy requirements, the inherent stability of the locomotion concept is another feature driving the control system. If a locomotion system can cope with a large variety of terrains without active control, the whole control system overhead can dramatically decrease.

The Solero concept follows this approach. A detailed description on the activity is given in [3]. It uses an innovative locomotion concept originating from EPFL, which is suited to provide a high degree of design–inherent stability with respect to locomotion in rough terrain.

F.6.1 Solero Requirements

The mission scenario for the Solero rover is regional exploration for a geochemistry mission. The payload for the SOLERO rover was chosen with reference to the *Nanokhod* microrover. It consists of an Alpha Particle X–Ray Spectrometer (APXS), a Mössbauer Spectrometer (MIMOS) and a small camera (MIROCAM). Accordingly, the top–level requirement for the rover system is to transport and operate these instruments for in–situ geochemical exploration.

The minimum on–surface travel distance for the Solero locomotion is specified to 10 km, with an effective locomotion speed of 220 m per day. In addition the rover system must be able to withstand Mars environmental conditions (e. g. outside ambient temperatures between $-100°$C and $+30°$C, Mars atmosphere, dust).

The Solero has a max. mass of 10 kg, an autonomous power supply using solar power collection, and –similar to the Nanokhod– has no active thermal control.

F.6.2 Solero flight concept for Mars

When defining a complete rover system flight concept, a variety of system elements have to be considered, defined and adjusted in an iterative and heuristic process. The design drivers for the Solero system design are:

- *Rover configuration, operations and control:* the rover must be able to carry out operations autonomously for a duration of at least one day. This implies the capabilities to autonomously solve the problems of rover localisation, path planning, and trajectory execution.
- *Power provision, storage and control:* the system has to work with a minimum electrical storage. As a consequence the diurnal power profile is driving the operational capabilities of the rover, i. e. driving, payload operation, telecommunication sessions, hibernation.

Further subsystems need of course to be assessed for the complete system concept, they are however not the strongest drivers.

F.6.3 The Solero chassis concept

Using a rhombus configuration, the rover has one wheel mounted on a fork in the front, one wheel in the rear and two bogies on each side. The parallel architecture

Figure F.2: The Breadboard of the Solero Minirover

of the bogies and the spring suspended fork provides a high ground clearance while keeping all 6 motorised wheels in ground contact. This ensures excellent climbing capabilities.

The front fork has two roles: its spring suspension guarantees optimal ground contact of all wheels at any time and its particular parallel mechanism produce a passive elevation of the front wheel if an obstacle is encountered. The front wheel has an instantaneous centre of rotation situated under the wheel axis that is helpful to get on an obstacle.

The bogies provide the lateral stability. To ensure similarly good ground clearance and climbing capabilities, the virtual centre of rotation of the bogie is set to the height of the wheel using a parallel configuration.

The steering of the rover is realised by synchronising the steering of the front and rear wheel and the speed difference of the bogie wheels. This allows for precise manoeuvres and even turning on the spot with minimum slip.

The payloads as well as rover subsystems can be accommodated in the central body of the rover. The flight concept of the Solero rover is equipped with a solar panel, scientific payload and two navigation instruments: an omni-directional camera and a stereovision camera for 3D obstacle recognition.

F.6.4 Conclusions for the Solero Minirover

The development of the Solero chassis was carried out within a very low cost technical demonstration activity (TDA), leading to a first system conceptual design and a development model (breadboard) to demonstrate aspects of locomotion, payload accommodation, and power provision. Although all system areas

Table F.2: The Solero flight concept is designed with the following technical attributes

Solero Properties	Target Value
Total Mass:	10 kg
Overall Dimension:	ca. $880 \times 600 \times 400$ mm
Payload Mass:	1 kg
Solar Power:	16 W typical daily peak power
Mean Locomotion Power:	8 W
Navigation:	Autonomous Navigation up to 1 km distance by 3D obstacle recognition and negotiation
Telemetry:	Close to the Beagle 2 design: direct telemetry link to orbiter total data size ca. 2 Mbit per day
Locomotion Speed:	20 cm/s

have been addressed in a first instance, some issues need however thorough design work in order to establish a detailed design suitable for a flight rover. The technical parameters of a Solero flight model are given in Table F.2.

This concerns in particular the control system. The current Solero model has only a very simple control functions implemented. It is therefore proposed to bring all subsystem areas to a detailed design level in a first development step called *Detailed System Design*. In a second step, the design, development and manufacturing of the EQM (Model) and the FM (Flight Model) rover could be implemented efficiently and with relatively low risk.

The positive results of the Solero activity recommend the Minirover concept as a promising solution for a future rover mission as well on Mars as on other planetary surfaces. It also recommends itself as a good base platform for various kinds of payloads which require an excellent mobility.

F.7 ExoMars Breadboard for European Mars mission

Bridget is the first full size ExoMars breadboard rover which was commissioned by Astrium UK Ltd. and built by a consortium of companies and institutes led by *von Hoerner & Sulger GmbH*, [4]. In the mean time the second ExoMars chassis breadboard was designed and built under the lead of vH&S together with institutes and the Swiss company Oerlikon Space.

The primarily intention of a full scale rover is to investigate the capabilities of the suspension and traction system for use on a future ExoMars rover. It is also used to study additional rover system components such as the navigation system or a drill. The scale of the rover chassis allows valuable experience to be gained not only in the performance of the systems but also from practical aspects such as accommodation of components and AIV aspects related to the handling of a full scale vehicle.

The breadboard chassis design reflects the shape and form of the proposed ExoMars rover at the time. However, pragmatic decisions in the use of materials and off–the–shelf hardware had to be made in order to keep to a reasonable cost for the rover whilst maintaining flexibility for its future use.

F.7.1 ExoMars Breadboard Design

The main design requirements of the rover chassis are based on the output from Astrium UK's Exomars/Pasteur Phase A mission study during which *von Hoerner & Sulger GmbH* had led the chassis study team.

The original phase A study proposed a 6 wheel rover with a RCL–C type configuration (based on the ESROL study) as the mission baseline, as it was a good compromise between complexity (and mass) and its performance in terms of stability and body movements during obstacle negotiation.

However, further comparison with miniature hardware rover models conducted for Astrium UK by ETHZ identified an undesirable characteristic which under certain conditions the outside wheels would effectively lift the centre wheel off the ground. A more complex control algorithm could be used to provide a solution to this problem but this is highly undesirable due to extra risk and resource it would entail. For this reasons plus the fact that RCL–E offers a reasonable mass advantage in a flight design, it was decided to proceed with a RCL–E type chassis configuration for the first breadboard design. This consists of a main body, two parallel bogies in the front and one lateral bogie in the rear

For the first breadboard the focus was placed onto an ideal passive suspension for obstacle negotiation in which there is only a minimum of longitudinal displacement of the wheels positions when the rover negotiates an obstacle. Longitudinal displacement of the six wheels relative to the centre of mass will cause the loads on each wheel to vary from the ideal situation where all wheels are equally loaded. This however, ignores the effect of the rover body inclination as

(a) Breadboard *Bridget* – 2006   (b) Breadbaord Phase B1 – 2008

Figure F.3: Development of two full-size ExoMars rover chassis Breadboards

it changes from horizontal. The longitudinal displacement was thus minimised for the Bridget Breadboard by optimising the geometry of the parallel bogies for the step range the rover had to cover, see figure F.3(a).

The drawback of such an ideal passive suspension unit is the additional mass, caused by the parallel beams. With the main focus on minimising mass for the second breadboard, the optimisation in the longitudinal displacement was given up. The new chassis was thus a simplified Concept E rover using simple bogies instead of parallel bogies. Only for the rear bogie the breadboard foresees the possibility to study the effects of both the parallel bogie and the simple bogie by applying or removing a locking device, see figure F.3(b).

During the Phase A Study of the rover chassis it had been highlighted that the use of a flexible wheel would be beneficial to vehicle performance and efficiency in several ways:

- for a properly designed flexible wheel, the larger (and longer) ground contact footprint will lead to less slip and higher thrust as compared to a similarly sized rigid wheel, resulting in better drawbar performance (and thus improved slope climbing capability)
- overall motion resistance of a properly designed flexible wheel is lower than that of a similarly sized rigid wheel, resulting in smaller losses or, equivalently, a better mileage (energy to be spent per distance driven).

This is why both breadboards have been equipped with flexible wheels.

For both breadboards, the main body accomplishes the chassis concept by

Table F.3: Main attributes of the ExoMars Breadboards

Properties	Breadboard Bridget	Breadboard Phase B1
Overcome obstacles up to:	0.3 m	0.25 m
Locomotion speed:	100 m/h	108 m/h
Power:	60 W peak, Battery	48 W Battery
Breadboard Mass:	~115 kg,	80 kg
Payload Capacity:	~185 kg	40 kg
Size:	1600×815×1000 mm	1650×1000×1200 mm

providing the linkage in–between the bogies. In both cases the body was made of profile frame in order to easily integrate additional systems and payloads.

F.7.2 Conclusions for the ExoMars Breadboards

Table F.3 compares the main parameters of the two ExoMars breadboards, which have been designed so far by vH&S. Since delivery Bridget has been used by Astrium (UK) for testing both in Tenerife and the UK, after the first preliminary tests at vH&S. The second breadboard is currently extensively tested at the company Oerlikon Space, where it was delivered to after completion of the assembly in Schwetzingen by vH&S. Valuable experience has already been gained during the project which will be put to good use for the next phases of the ExoMars project. Further field trials and performance test undertaken will build further on this experience. Both rover have appeared extensively in press and television, being the first visible steps of the European rover mission to Mars.

F.8 Conclusion

The development and design of rover is a multi–disciplinary task due to the complexity of a space system. Such a system has one common goal, which is to transport and operate scientific instruments on a planetary surface in order to collect data and provide knowledge of our solar system. Nevertheless the space system rover consists of several different components, like drive units, the control unit, the chassis etc., which can be defined as subsystems. These subsystems have again their own properties and functions and in–between them there are a high number of interactions and exchanges. This interaction between the different elements requires an overall design, which allows the integration of all elements and subsystems and thus maximising the possible synergies. In addition, the definition of such a subsystem as well as the input and output values can change depending on the applied point of view.

Applying to a space system additional requirements imposed by the mission scenario, the encountered environment or the influences of the rover size onto the chassis, it becomes obvious that the development of a matching rover is not only the application of size factor.

Quite the contrary, each rover design has to be considered as a "network–type" problem. For all missions a new set of requirements has to be developed and designed to. This means the design follows a heuristic, non–sequential method-ology to meet the demanding requirements any space mission for planetary exploration imposes.

Nevertheless vH&S has proven a large experience in this field with the rover designs on hand. The company has solutions available for issues concerning small–scale systems, Minirover scale as well as large rover systems, see figure F.4.

Figure F.4: The magnitude variety of vH&S rover

F.9 Bibliography

1. Klinkner, S., Laufer, R., Graf, T., Nagy, M., Roeser, H.-P., *Lunar Exploration using Small Satellite with Micro Rover Technology*, 10th ISU Annual International Symposium, Strasbourg, France, (2005)

2. Schiele, A., Romstedt, J., Lee, C., Henkel, H., Klinkner, S., Rieder, R., Gellert, R., Klingelhöfer, G.,Bernhardt, B., Michaelis, H., *The new NANOKHOD: Engineering model for extreme cold environments*, 8th International Symposium on Artificial Intelligence, Robotics and Automation in Space, September 5 – 8, 2005, Munich, (2005)

3. Michaud, S., Schneider, A., Bertrand, R., Lamon, P., Siegwart, R., Van Winnendael, M., Schiele, A., *SOLERO: Solar Powered Exploration Rover* ASTRA 2002, Nordwijk, (Nov. 2002)

4. Lee, C G–Y., Dalcolmo, J., Klinkner, S., Richter, L., Terrien, G., Krebs, A., Siegwart, R., Waugh, L., Draper, C., *Design and manufacture of a full size Breadboard ExoMars Rover chassis* 2006, Nordwijk, (Nov. 2006)

G Astra Conference Paper 2008

10th ESA Workshop on
Advanced Space Technologies for
Robotics and Automation

EXPERIENCES GAINED FROM THE THERMAL VACUUM TESTS OF THE MICROROVER NANOKHOD

Authors: Sabine Klinkner[1,2], Chris Lee[1], Carsten Wagner[1],
Michael Lengowski[2], Hans–Peter Röser[2], Pascale Bourlier[3]

Astra 2008

11. – 13. November 2008, at ESTEC
Nordwijk, the Netherlands

[1] *von Hoerner & Sulger GmbH,*
Schlossplatz 8, 68723 Schwetzingen, Germany,
Email: klinkner@vh–s.de

[2] *Universität Stuttgart, Institut für Raumfahrtsysteme,*
Stuttgart, Germany

[3] *Harmonic Drive AG,*
Limburg, Germany

G.1 Abstract

With the recent development of the microrover Nanokhod for the demanding surface environment of the Mercury night side, the rover made a huge advance towards a practical flight model. The rover was designed by the company *von Hoerner & Sulger GmbH* under an ESA contract to meet the flight–model requirements for a Mercury surface mission. This means it has to withstand such extreme environmental settings as temperatures down to $-180°C$ in vacuum conditions, a landing shock of 200 g for a duration of 20 ms and a surface material consisting of very fine regolith with an extremely low thermal conductivity. The hardware model based on these requirements was manufactured to an engineering level and is suitable for environmental testing of vibration, shock and thermal vacuum.

The first thermal vacuum tests of the rover system were successfully conducted and the results provide a good understanding of the thermal behaviours within the system. In addition, these results are used for the validation of the detailed thermal model of the rover which has been developed in parallel to the Nanokhod model. This validated model thus presents a powerful tool which can be used as a baseline for the future design steps of the thermal concept. It provides detailed information on possible hot spots in vacuum condition and helps to avoid them. A good understanding on the temperature distribution within the rover also helps to reduce the risks and uncertainties for the scientific instruments and critical electronic components.

In addition to the first environmental tests, the drive units of the rover system were also subject of further analysis. For the new rover development the drive unit had to be redesigned to match the environmental requirements of the Mercury surface. Unfortunately the framework of the project did not allow for a workbench setup of the drive system before the integration into the small scale rover volume. This lead to some initial blocking problems. These problems could be solved by a run–in and some adjustments of the drive; the rover now works fine. Nevertheless, for the benefit of a future flight model the drives have been thoroughly tested to have a good understanding of the different factors having an influence on the drive unit and then to be able to provide a reliable drive system.

Although currently the mission to Mercury is very unlikely, the new design prepares the rover well for a number of alternative mission scenarios. Especially the detailed analysis of the thermal aspects of such a small and integrated system offers a perfect baseline for mass critical and thermally challenging requirements as for example on a mission to the moon poles.

This paper describes the results of the thermal vacuum tests of the rover as well as the lessons learnt on the detailed drive analysis.

G.2 Introduction

The development history of the microrover Nanokhod under the leading role of *von Hoerner & Sulger GmbH* started in the early 1990's. Since then the original Russian concept has been adapted for several target surfaces like the Moon, Mars and Mercury. In the development for example, a package of three scientific instruments was accommodated into the rover. The current rover payload consists of two spectrometers – an Alpha–Particle X–ray Spectrometer and a Mössbauer–Spectrometer – and a Micro–imager. The development also included the building of microrover prototypes which were used to test the rover abilities, for example the driving performance on soil simulants.

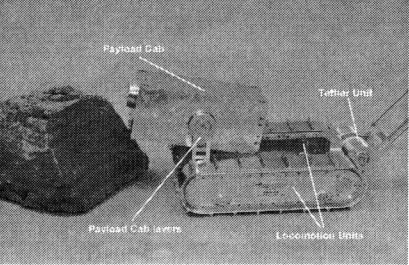

(a) For size comparison of the microrover   (b) Naming convention of the Nanokhod

Figure G.1: MRP mircrorover Nanokhod

The Nanokhod rover is a small mobile scientific platform, designed to transport and operate scientific instruments for in–situ measurements of rocks and small craters in the vicinity of the landing point. The microrover has a volume of $160 \times 65 \times 250\,\text{mm}^3$, it weighs 3.2 kg including a payload mass of 1 kg and it has a peak power of 5 W. The Nanokhod is a tethered system that uses the Lander for power supply and as its data relay to Earth. An impression of the rover size and the labelling of the main rover components are given in Figure G.1.

In the course of the latest development the rover gained a near flight readiness with a practical design to withstand the tough requirements of a flight model for the Mercury surface [1]. Having realised a design that meets the flight–model requirements, a hardware model has been manufactured to an engineering level which is suitable for environmental testing of vibration, shock and thermal vacuum. This design is able to withstand such extreme requirements as temperatures down to $-180°C$ in vacuum conditions, a landing shock of 200 g for a duration of 20 ms and a surface material consisting of very fine regolith with an extremely low thermal conductivity.

Due to limited resources and delays within the original TRP study, the environmental tests of the model were discarded from the project. To still validate the rover model, further analysis and tests were conducted in the framework of an University project together with the IRS (Institute of space systems, Universität Stuttgart). These analyses included the development of a detailed thermal model of the rover and the first thermal vacuum test with a moderated thermal range. The results from the thermal model were used to optimise and validate the thermal model.

Further tests considered the drive unit of the rover. During the initial functional tests of the manufactured model, the drive unit caused some problems. Although the problem could extensively solved during the project, some further drive tests were carried out after the completion of the MRP project, to fully understand the problem. These tests analysed the effects of the drive unit in more detail and thus help to provide a reliable drive unit for any future implementation of the microrover.

G.3 THERMAL MODEL AND THERMAL VACUUM TESTING

For a complex system like the rover, a simple thermal estimation will give some first numbers and ideas in which dimension the temperatures will range for the required environmental conditions. However, in order to get representative limit values for the electronic parts or the scientific instruments, it is essential to model the rover system and the occurring environment in more details.

In parallel to the development phase of the rover the thermal model was build and refined to more and more detail. The first model was a simple spread sheet calculation giving temperatures in rough approximation. The second model is designed with the help of a Design Compiler 43 and the ESATAN–/ ESARAD– Software. Based on the second model there has been a number of iterations on the thermal network considering areas of special interest and refining the network due to unexpected temperature measurements or on thermally sensible parts.

The rover breadboard was designed and built for testing purposes in a thermal vacuum chamber for the boundary conditions of the Mercury surface. This is why the thermal rover model was reconfigured to provide the boundary conditions of thermal chamber. The thermal model could thus be used to predict the temperatures for different conditions in preparation of the thermal vacuum tests. Additionally the modification of the model allowed using the test results to evaluate the model. Deviations between the temperature results of the test and the model are caused by the assumptions made on material properties and simplifications in the model, which can be improved or even completely resolved.

A thermal model calculation before the first thermal vacuum test, assuming the vacuum test boundary conditions of the thermal vacuum chamber, was conducted to show that the planned functional rover tests can be safely accomplished also for room temperature conditions. The model showed that the temperatures of the heat dissipating electronics would not exceed critical levels and damage could occur.

G.3.1 Objectives of thermal vacuum tests

The main objective of the thermal vacuum test is to verify the compatibility of the MRP design to the thermal vacuum environment. Although the facility used for the thermal vacuum testing was not able to reach the required temperature of $-180°C$, it allowed to test the system on relieved temperature levels, as a first qualification step of the rover system. Furthermore, the test provided data for the comparison with the calculations of the thermal model realised with ESARAD/–TAN.

There are two separate thermal vacuum tests in order to gain as much information as possible from the rover. The two tests allow a higher number of test data, as well as the possibility to have a closer look at specific temperature distributions during the second test after the evaluation of the first test data. Part of both thermal vacuum tests was a rover functional test for the verification of the rover system. The functional test involved the operation of the electronic system and the PLC mechanisms with the rover static on its tracks.

G.3.2 Comparison of test results and iterated thermal modelling results

The calculated temperatures of the thermal model were compared with data of each test. The comparison helped to optimise the thermal model by modifying some simplifications and assumptions, in order to achieve matching temperature curves. The results from the optimised model compared to the test data of the second thermal test show finally a good compliance. Still the temperature curves from the calculated and the measured data do not cover each other completely, due to simplifications within the model and effects of the temperature measurement. For example, one simplification effect in the model is, that the calculated temperatures are always an average value for the whole component. The effect is larger for components, which partially dissipate heat like the PCBs, due to the higher thermal gradient within the component. Also large components show a higher inaccuracy as the average temperature is taken over a larger volume.

Measurement divergences result from the fact that the temperatures are only determined on one surface of the component and thus depending from the conductivity within the component, the measured value will give only local information. Additionally the mounting of the thermistors on the surfaces has to consider the impact of thermal radiation effects from the surrounding components. The effect of the thermal radiation on the thermistor is quite high due to its low heat capacity in comparison to the component on which it is mounted.

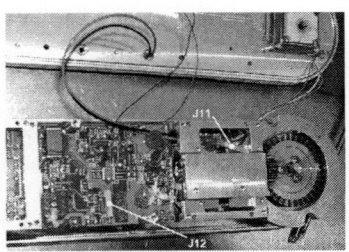
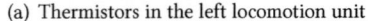

(a) Thermistors in the left locomotion unit   (b) Thermistors on the left locomotion unit

Figure G.2: Positioning of the thermistors on and inside the rover during thermal vacuum test

Figure G.3 and G.4 show the comparison of measured and calculated data on two heat dissipating and two purely passive component examples (figure G.2).

The two heat dissipating components are inside the left locomotion unit (LU): The motor controller and the Payload Cabin (PLC) articulation motor located in the front of the left LU. The thermistors are indicated with J11 (motor) and

J12 (controller PCB) in figure G.2(a). The controller PCB is inside the track unit in-between the two side walls, to which it is mounted via aluminium standoffs. The electronics on the motor controller PCB are activated approximately half way through the functional test when the articulation motor in the locomotion unit is switched on.

The thermistors on the two passive components are sensor J7 and J8 on the track foil on the upper and lower side of the left locomotion unit, see figure G.2(b).

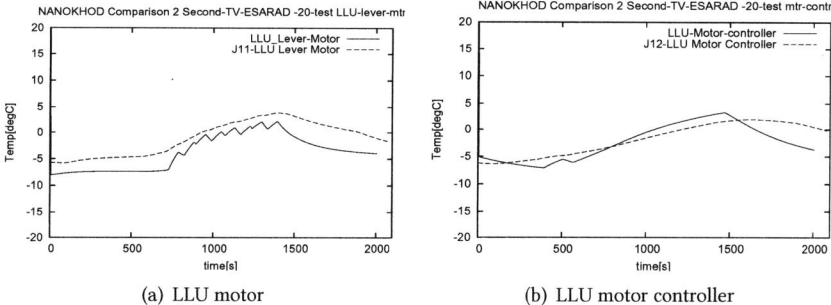

(a) LLU motor   (b) LLU motor controller

Figure G.3: Comparison of the validated model results and temperatures of the second thermal vacuum test for the LLU motor components

Figure G.3(b) shows a good compliance of the calculated and measured temperature of the motor controller. The calculated temperature shows a very distinctive curve, while the measured curve is following the same trend with a smoother line. The sensor is measuring a smoother temperature behaviour, as it is on the surface of the motor controller PCB, and thus the dissipated heat which is reaching the sensor is slightly damped. For the calculation the heat is dissipated directly within the component and thus the temperature curve follows the activation of the motor controller without delay.

The sensor directly on the articulation motor in the LU (figure G.3(a)) shows that the calculated temperature matches nicely the measured temperature curve; however, the measured data is again less distinctive due to the position of the sensor. The sensor is mounted on the outside of the motor cradle where the maximum thermal resistance of the part applies. Opposed to the thermal resistance assumed in the calculation which is an average of the thermal resistance in the centre of the motor to the maximum thermal resistance on the surface of the cradle. At the same time the thermal radiation effects of the cold LU parts, which are surrounding the motor, affect the measurement of the thermistor. Thermal radiation from the motor controller also explains the slightly rising temperature value, before the articulation drive is activated.

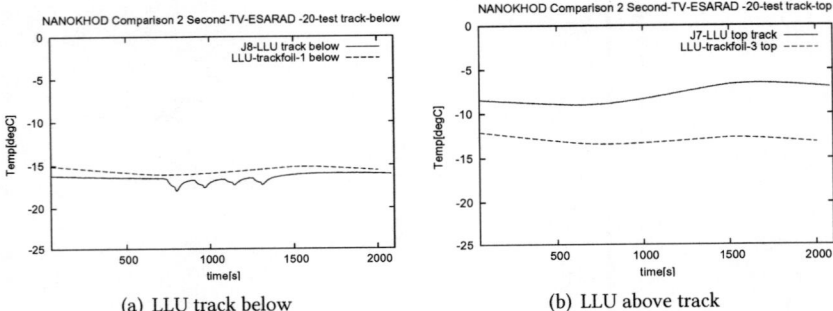

(a) LLU track below   (b) LLU above track

Figure G.4: Comparison of the validated model results and temperatures of the second thermal vacuum test for the LU track foils

Graph G.4(a) in Figure G.4 shows that the curves for the lower side of the track foil match quite nicely. The only differences are some waves in the measured curve during the activation of the articulation drive. These result from the rotation of the drive, providing a changed radiation interface to the sensor by moving a colder surface of the drive towards and then away from the sensor. The movement of the surface is not included in the thermal model and thus can not be seen in the calculated curve.

For the upper side of the track foil, the calculated and measured curve match very good (figure G.4(b)), however the measured curve is always higher than the calculated one. This is caused by the assumption within the thermal model that the boundary nodes, a cooling plate below the rover and a shroud covering the rover, have each one temperature across their complete surface. However, the boundary nodes in the test have a gradient; the shielded area underneath the rover is much colder than the area surrounding the rover. Thus the model assumes a too low temperature for the area around the rover, when using an average temperature and on the opposite a too high temperature for the shielded area underneath the rover. Due to that the calculated temperature for upper side of the track foil is lower than the measured one.

Despite the slight deviations from the calculated and the measured temperature the model shows a good compliance with the real conditions in the test. All observed variations are marginal and can easily be explained by either the thermistor position or a simplification within the model.

G.3.3 Influence on the thermal system design

The iterated model can now provide the temperature distribution of all rover components which could not be measured due to the limited amount of thermistors or a limited accessibility. Additionally the model can be used to implement different boundary conditions, in order to analyse the thermal behaviour of the rover within any implemented environment. Like that the model can disclose critical areas for the applied environmental conditions and allows the optimisation of the thermal system concept in case it becomes necessary. With these results the thermal system of the rover can be validated for the target environment of any mission.

The last iteration of the model implements the boundary conditions of the Mercury mission scenario for the BepiColombo lander. The contact surface is assigned with a surface temperature of $-180°C$ and with a thermal conductivity for regolith of $\lambda = 0.01\,\text{W/mK}$ [61].

A calculation was conducted based on the deployment and operation procedure for the first 48 h as foreseen for the BepiColombo lander mission. The scenario consists of the following rover activities, including the mentioned rover modes:

- **Checkout sequence**, including the rover modes Localisation, Science APXS, Science MIMOS, Science MIROCAM
- **Complete Path**, including the rover modes Locomotion, PLC Move, Localisation, Locomotion, Localisation
- **Measurement Sequence**, including the rover modes Locomotion, PLC Move, Science MIROCAM, PLC Move, Science APXS, PLC Move, Science MIMOS, PLC Move
- **RoverOff**

Figure G.5 shows the temperature distribution within the rover after a Complete–Path activity (figure G.5(b)) as well as after a Measurement–Cycle activity (Figure G.5(a)).

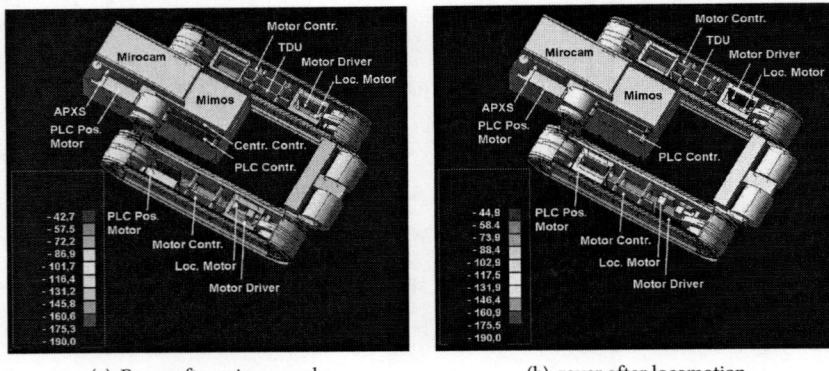

(a) Rover after science cycle   (b) rover after locomotion

Figure G.5: Temperatures in the rover after science cycle and locomotion activity

In both cases the temperature within the rover varies between approx. $-44°$C on the hottest component and $-180°$C on the Mercury surface node. During the scientific measurements the highest temperature can be observed on the central controller PCB in the PLC. Also the PLC motor is quite warm due to the repetitive PLC rotation for the positioning of the instruments. And as the last instrument of the measurement cycle the MIMOS instrument is still on a temperature level of approx. $-70°$C.

For the locomotion mainly two areas have an increased temperature: The locomotion driver PCBs in the rear of each locomotion unit and slightly less the central controller PCB in the PLC. As shown on these two examples the model provides the thermal distribution within the system for any rover mode. Like that it defines the thermal boundary conditions which have to be applied for each component and thus supports any future design.

G.4 DRIVE TESTING

Tests on the drives system have clearly demonstrated that it is capable of producing sufficient torque for the application. However, during the initial testing of the rover the drive occasionally stalled whilst running. Investigations indicated that a number of factors are causing the torque of the stepper motor to be reduced below the required limit. The variation in the drive torque can cause the motor to miss a step. The problem is that a stepper motor is unable to run immediately at the operational speed and requires a controlled acceleration ramp, which means if the motor misses a step due to a torque peak, it will stall. As the stepper motor is driven with an open loop control the motor does not recover autonomously from the this situation. Several examinations have been carried out, when the stalling of the drive occurred during the functional tests of the rover, with the following results:

- Software/electrical drive problems are discounted after tests.
- Stalling position suggests no single tooth of crown/pinion gear is responsible for the stopping of the motor.
- Load is a factor in the frequency of occurrence.
- The problem occurs more often in one direction of output shaft (but opposite for the two PLC drives). The crown gear is mounted in different orientation for the PLC− (teeth towards HD) and the Lever−drive (teeth away from HD). This means that the same crown gear face is causing the higher probability of errors. This can be an indication for a machining error.

Motor tests on a spare motor showed that the clamping of the motor body caused the motor to stop, but also that it is capable to produce more than the required torque. The re−testing of the drives while improving the method of clamping the motor showed some improvement. Another improvement could be achieved by enlarging the distance between pinion and the gear wheel of the crown gear to a higher value than required by the gear setup tolerances. The bigger distance reduced the radial loads on the planetary gear axis which have caused a reduction of torque.

Still the problem could not be completely solved and the stalling kept occurring occasionally. However the problem appeared less frequently the longer the drives were running. Some running−in effect seemed to improve the situation. This fact correlates very well with the friction characteristics of Dicronite lubricant over the number of revolutions [2]: The friction is approximately 3 to 4 times higher for the first revolutions, after a running−in time it is dropping to the low friction coefficient of 0.03, before it is suddenly rising excessively at the end of its lifetime.

The mounting tolerances over the whole drive system interfaces are another probable factors leading to the reduction of torque margin. The Harmonic Drive unit comprises already nine interfaces. Also the dry lubricant may increase the stiction forces of the faces of the gear teeth. All effects are exacerbated by the small scale of the components relative to the used torque levels. Although the problem seemed to disappear after the running–in of the system, the drive was carefully examined in two tests:

The tests were carried out at the Harmonic Drive AG laboratory and were considering the validation of the drive system and its margin. The objective of the first test was to inspect the MRP drive unit (consisting of the planetary gearhead, the crown gear and the Harmonic Drive unit (HFUC–5–100) driven by the stepper motor AM1020), when mounted in a stiff environment on the test bench. The test setup provided assessments of torque variations on the output as well as acceleration measurements. The objective of the second test was to verify the torque loads specifically in the Harmonic Drive stage during the revolutions of the gear. The second drive setup was done with a DC motor in order to monitor the current which directly relates to the required torques.

G.4.1 MRP drive unit test with stepper motor

For the test a 50 cm arm is attached to Harmonic Drive flexspline output of the LU lever drive to apply torque to the drive. The arm has a bucket on the other end to add weight to adjust the torque on the drive. The arm starts from vertical downwards position and moves counter–clockwise to the horizontal position. With an unloaded bucket the maximum torque in the horizontal position is 0.35 Nm. By adding weight in steps of 60 g the torque applied to the drive is raised to max. 2.7 Nm. For each load step the arm is driven counter–clockwise from the vertical to the horizontal and back down again. The test is repeated with three different step rates: the nominal step rate of 1528 steps/s, the step rate of 934 steps/s and the slowest step rate of 104 steps/s. The test–setup is shown in figure G.6.

The inductance of the motor windings of a stepper motor changes when load is applied [3]. In–between two steps of the motor the application of the voltage moves from one coil to the next. The so called rise time that is needed for the current to build up in the new coil and is dependent from its inductance. With a higher load the inductance decreases and the rise time is shorter. Thus the measurement of the rise time indicates the torque load applied to the motor. For the detection of torque variations during the test, an oscilloscope measures the voltage and the current and gives a trend of the rise time over a period of 20 sec.

During the tests no stalling of the drive system occurred, nevertheless the

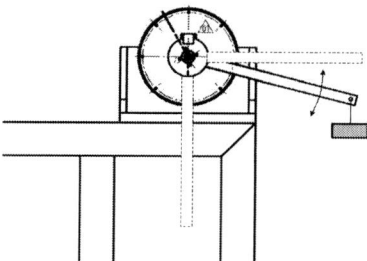

Figure G.6: Test setup of drive tests

measurements of the first test confirmed the suspicions for causes of stalling. The measurement results for the test with minimum and maximum torque load and for the nominal step rate are shown in the figure G.7. The graphs present the rise time for the switching of the motor coils in–between two steps applied over the time to lift the arm from vertical to horizontal. The graphs show a clear trend – the green line –, derived from an average over 40 values, of decreasing rise time and thus increasing torque, for the loads applied. This trend becomes more obvious when applying a higher load. Spikes of shorter rise time in both graphs indicate several torque peaks during the lifting. This explains the higher occurrence of the problem with higher torque loads, as the torque peaks add to the general load and thus can lead to torque levels which are larger than the motor margin allows for.

An acceleration sensor was attached to the output of the drive. At first sight, its measurements reveal no influence of the different gear stages as the results are too noisy. In order to see if the data is a combination of sinusoidal basis functions which identify specific influences, a Fast Fourier Transformation (FFT) analysis is carried out on the data of the acceleration sensor. The comparison of the FFT analysis for the nominal step rate with medium and maximum load applied to the torque arm (figure G.8), shows the noisiness of the acceleration data. However, for both torque loads the same two frequencies show the highest amplitude, in the range of ~ 22 Hz and in the range of ~ 47 Hz.

The amplitude of these frequencies varies with the torque load. Tests with slower steprates revealed, that the system shows the same peaks, even though the stimulated frequencies of the different gear stages are lower for the slower speeds. The amplitude of the frequencies is the only parameter that varies with

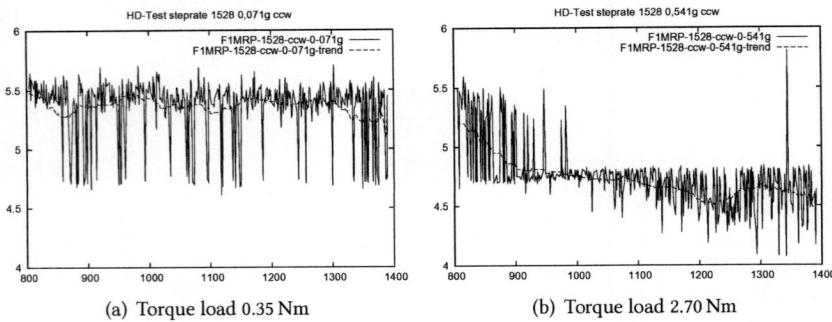

(a) Torque load 0.35 Nm   (b) Torque load 2.70 Nm

Figure G.7: Rise time distribution over a 90° counter-clockwise output movement for different loads and the nominal step rate of 1528 steps/s

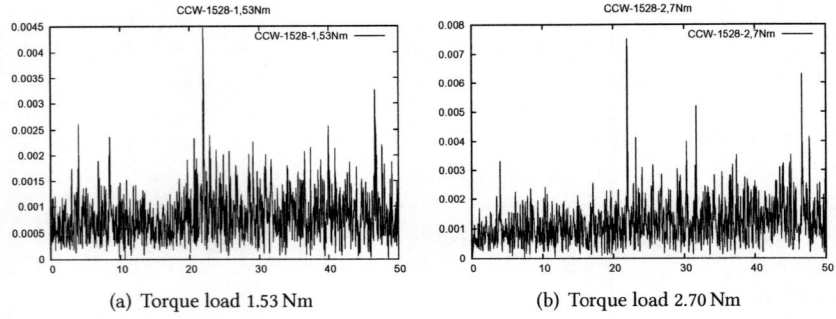

(a) Torque load 1.53 Nm   (b) Torque load 2.70 Nm

Figure G.8: FFT analysis for stepper–drive tests with different applied torque loads and the nominal step rate of 1528 steps/sec

the motor speed and the torque load. The fact that the frequency is the same for all motor speeds implies that this is the Eigen–frequency of the gear unit setup. A comparison with the frequencies of the different gear stages in table G.1 does not indicate that the Eigen–frequency is driven by an irregularity of one of the gear stages.

However the impact of the gear–stages can not be completely disregarded as it might be covered by other effects. Unfortunately the MRP project did not allow for careful testing of the separate gear stages before assembling the unit and integrate it into the rover. A following phase should include a comprehensive test campaign, which is analysing the single drive stages as well as the unit. Only the proper analysis of the drive unit will provide a reliable drive system the rover can rely on.

Gear stages	Step rate 1528 steps/s		Step rate 934 steps/s		Step rate 104 steps/s	
	Angular frequency [rad/s]	Frequency [Hz]	Angular frequency [rad/s]	Frequency [Hz]	Angular frequency [rad/s]	Frequency [Hz]
Harmonic Drive	0.49	0.08	0.30	0.05	0.03	0.005
Crown gear	1.19	0.19	0.73	0.12	0.08	0.013
Planetary Gearhead	76.34	12.15	46.70	7.43	5.20	0.828

Table G.1: Frequencies of the drive stages for the different motor speeds

G.4.2 MRP drive unit test with DC–motor

The objective of the second test is to verify the torque loads specifically in the Harmonic Drive gear stage. In this test the Drive unit is disassembled and only the Harmonic Drive stage is setup for the test and driven by a DC motor. The application of torque loads is done in the same way as for the stepper motor tests: The drive is mounted onto a rigid test bench. A load arm is applied to the output of the Harmonic Drive. The arm is driven counter–clockwise from the vertical to the horizontal position with different loads added on the other end of the arm, providing an increasing torque for the lifting of the arm. Measurement devices record the input torque needed to back–drive the gear. The test shall verify if the torque peaks which were observed in the stepper motor setup can be detected also in the Harmonic Drive stage. The input speed on the Harmonic Drive input side is 26 rpm.

The graphs in figure G.9 show the back–drive torque for two different load cases. As expected, the trend (derived from an average over 40 values) shown as a green line of the back–drive torque is deflected from the zero axis for an increasing load of the arm travelling from vertical to the horizontal position.

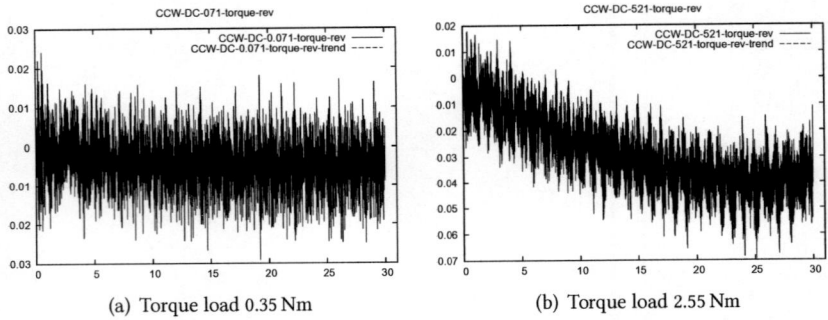

(a) Torque load 0.35 Nm   (b) Torque load 2.55 Nm

Figure G.9: Back–drive torque over a 90° counter–clockwise output movement for different loads

This effect becomes more obvious for an increasing torque load. Torque peaks as for the stepper motor tests can not be detected. However, the measured data is relatively noisy resulting from the sensitive drive control. This means possible peaks could be covered by the noise. The separate test of the Harmonic Drive unit does not indicate any irregular running of the Harmonic Drive.

The Eigen–frequencies derived from the FFT of the acceleration data and shown in figure G.10 do not correspond with the frequencies of the drive unit rotation, given in table G.2. Due to the disintegration of the drive unit, the Eigen–frequencies for the DC test differ from the Eigen–frequencies of the stepper motor setup. The impact of the different applied torque loads becomes again obvious in the variation of the frequency amplitudes.

Gear stages	DC	
	Angular frequency [rad/s]	Frequency [Hz]
PMA	0.43	0.07
Harmonic Drive	43.35	6.90

Table G.2: Frequencies of the Harmonic Drive for the DC motor test

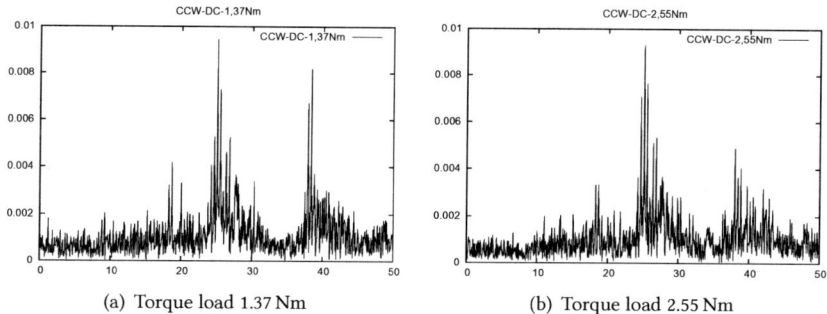

(a) Torque load 1.37 Nm   (b) Torque load 2.55 Nm

Figure G.10: FFT analysis of the acceleration data for PMA–drive tests with different applied torque loads

G.4.3 Thermal test of an alternative actuator – the brushless EC10

In the course of the project an alternative actuator came on the market. The brushless EC10 is of a similar size than the currently applied motor AM1020 and it is able to provide a higher torque as well as a higher output speed of the rover. In the framework of the drive system analysis a cold test of this motor was conducted. This test was conducted to verify the function of the brushless motor for the temperature range down to $-180°$C.

The first test with a sensorless motor showed that the motor works down to $-180°$C. Further tests with a brushless EC10 including a hall–sensor showed that the motor still works reliably to the low temperatures, however the hall sensor signal fails. The signals started disappearing at a temperature of approx. $-80°$C.

For the purpose of using the brushless DC motor for the rover drive design the encoder unit of the motor has to be reviewed. In order to have the brushless DC motor as a promising alternative actuator the hall sensor could be replaced with another encoder type, which works reliably down to $-180°$C.

The implementation of the brushless DC–motor with a customised encoder, allows for a bigger torque margin, resulting into a more reliable drive unit.

G.5 Conclusion

The development of the miniaturised robotic space system Nanokhod made a big step forward in direction of the flight model. On the mechanical side the

Nanokhod rover was developed from a concept stage to an engineering model, which is able to withstand the demanding conditions of a Mercury mission. Additionally the new rover design was optimised concerning the critical resources of volume, mass and power. The current design represents a highly integrated, mobile payload which can be flexibly implemented on a variety of missions for the exploration of planetary surfaces.

The extreme environment made it necessary to develop a more sophisticated thermal rover model. The development of the model was optimised and validated in the first two thermal vacuum tests of the rover systems. The validated model can now be used to predict temperatures of specific components. This makes the choice of components less risky and thus less cost intensive when carrying out the next design steps. The thermal model additionally allows an easy reassessment of the thermal design, e. g. for alternative mission scenarios.

On the mechanical side the solution of a reliable drive implementation was analysed to far extends. However the next design step shall foresee further work on that issue: The development of a suitable encoder for the brushless DC motor solution should be included. It is essential to qualify the drive components separately from the rover on an equivalent test bench which allows detailed measurement of the performances. This ensures that all contributing factors can be fully identified before implementing them in the complete drive unit. The test described in this paper provide already valuable experiences and thus provide a good baseline for further tests and for a reliable drive unit solution.

The rover is at a high state of development; although no mission opportunities currently exist, it is expected to be seriously considered for future missions to the Moon, Mars and other destinations.

G.6 REFERENCES

1. Klinkner, S., et al, *Destination Moon and beyond for the Microrover Nanokhod*, DGLR International Symposium, Bremen, Germany, Mar. 2007

2. Anderson, M.J., Cropper, M., Roberts, E.W., *The Tribological Characteristics of Dicronite*, ESTL, ESR Technology Ltd. UK, 2007

3. Schlienz, U., *Der Schrittmotor der nicht ausrastet*, WEKA FACHMEDIEN GmbH, 2007

4. Barbé, J., *Mercury Thermal Model*, ESTEC–Noordwijk, (Aug. 2003)

H ISU International Symposium Paper 2005

Lunar Exploration Using Small Satellite with Micro Rover Technology

Authors: S. Klinkner[1,2], Rene Laufer[2], Tanya Graf[2], Michael Nagy[2] and Hans-Peter Roeser[2]

10th ISU Annual International Symposium 2005, Strasbourg
Space Exploration: Who, What, When, Where, Why?

[1] von Hoerner & Sulger GmbH, Schlossplatz 8, 68723 Schwetzingen, Germany,
Email: klinkner@vh-s.de
[2] Universität Stuttgart, Institut für Raumfahrtsysteme, Stuttgart, Germany

H.1 Abstract

Current extraterrestrial science and exploration preparation missions require a significant augmentation of the payload mass to resource ratio. On the one hand this is due to the energetic more demanding missions and on the other hand the financial resources are more and more limited and thus do not allow large-scale explorations, which would be for example launched with more than one launch system. Nevertheless, mobility as well as a larger number of probes and stations is a key factor in order to achieve a wide spectrum of science exploration in the orbit and on the surface of the target planet.

With small satellites dedicated scientific instruments or surface stations can be delivered into specific orbits or to specific surface locations. By using a mobile research platform, the scientific instruments can be carried directly to the point of interest, for example rocks and craters, in the near surrounding of the landing point. Especially small low–cost missions like LUNAR MISSION BW1 of University of Stuttgart with only low of number of instruments are able to fulfil specific orbit demands without compromises. With these system drivers in mind, small satellites and microrover technology represent an optimal solution to keep the overall system mass relatively low and still have the opportunity to perform remote sensing tasks and examine the wider surroundings of a landing site.

This additional flexibility of a microrover relative to other landing systems and robotic arms allows the mission scenario to be customised to the landing area that is found. Furthermore the ratio of scientific payload to resource mass for microrover can be very high as is demonstrated for example by the NANOKHOD of *von Hoerner & Sulger*, Schwetzingen, Germany. Taking this into account the possibility of a lunar NANOKHOD mission in combination with the LUNAR MISSION BW1 was studied. Results of the study made in cooperation between the *Institute of Space Systems* of the *University of Stuttgart* and *von Hoerner & Sulger* will be presented.

H.2 Introduction

Since the beginning of space exploration, the limitation of mass has been one of the strongest drivers and still the need for increasingly payload mass to resource ratios remains extremely high.

In order to achieve a significant augmentation of the payload mass without the need for more resources, a full range of issues have to be considered due to the complexity of a space system. In any case, mass reduction and miniaturisation have to be promoted in all disciplines which participate in a space system. However in addition to all these efforts in the individual disciplines, the main challenge is considering and dealing with the whole system as a whole and its internal interactions. This is particularly true for small space systems, which require greater use of highly integrated structures to achieve their goals. This paper discusses the necessity of utilising highly integrated design techniques in a small space system using a Micro rover (Nanokhod), which is designated as a payload for a small satellite (Lunar Mission BW1) as an example.

H.3 Small Space Systems

Small space systems have become increasingly necessary in today's space exploration. This is due to several different reasons:

First of all the financial resources are much more restricted then in the first decades of space science. This is because the strong political driver (e. g. cold war), that pushed space science at the beginning, has decreased in the last years. In addition to this, the spectrum of space science has recently become a lot wider, so that the overall budget has to be divided into a higher number of shares. This is especially obvious in the European space industry; the interest of the different participating countries in space science varies strongly, and in order to fulfil all the different claims, the European space industry undertakes a large variety of missions, though all with a quite restricted budget.

A further driver for small space systems is the extension of the exploration horizon. Expanding the field of interest from the near Earth orbit first to the planets like Mars and Mercury, and than to further targets like their moons and the outer planets is energetically much more demanding and requires a greater mass of propellant. Increasing the propellant mass, implies a reduction of the payload mass, if the overall mission envelope is not extended for example by using a higher quantity of launchers.

All these reasons emphasise the need for small systems in future space missions.

H.3.1 Definition of System in General

The following attributes define a system:

- It has a common goal / objective
- It consists of several elements with attributes, functions, inputs and outputs
- There is an interaction between these elements
- The system has a hierarchical structure
- The definition of the elements, borders, inputs and outputs may change with the viewpoint

H.3.2 Definition of a Small System

The following points are added in, in the particular case of a small space system.

- Low budget solution
- Shorter development period
- A small number of instruments capable of fulfilling specific orbit demands without compromises
- Highly integrated system/subsystems

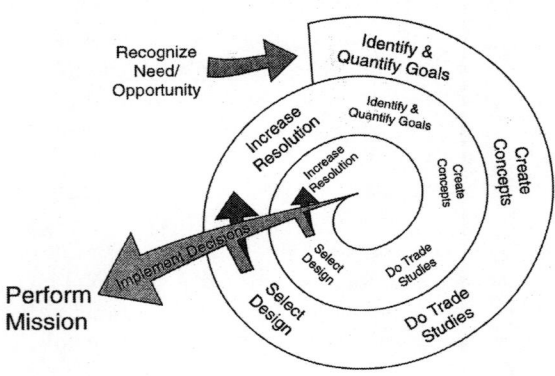

Figure H.1: System Design Scheme

It can be seen from these definitions that the improvement of the payload mass to resource ratio may only occur by a broad study over the whole system, due to the interactions between the individual elements. The bettering of the ratio, without compromising the scientific performance, implies a system wise approach (compare also Figure H.1).

First of all, investigations and efforts have to done in all subset of the system design in order to gain mass reduction and miniaturisation in every field. This can be the development of lighter but stronger materials, as well as the miniaturisation of electronics or scientific instruments. However all these efforts will only have a limited success, in relation to the whole system. In addition to considering the whole system, better results can be achieved, by using all possible synergies already at the lowest level of space system design. This means moving away from a design of discrete boxes.

Especially in small space systems the need of highly integrated structures is evident, as single items of the system take a much larger proportion of the system resources, then the same item in a bigger system. Thus a greater proportion of mass and volume can be saved, by removing the item in a highly integrated system.

This interaction between the different elements requires an overall design, which allows the highly integration of all the elements and subsystems and thus maximising the possible synergies.

This is done by:

- Considering the design as a "network–type" problem
- Developing requirements and design together
- A Multi–disciplinary approach of a methodological problem
- Heuristic, non–sequential, baselines & iterations a clear methodology and adapted tools

H.4 Mobility in Space

Current trends not only aim for smaller systems but also for mobile systems. In order to fulfil the whole range of scientific interests, the instruments exploring the surfaces of other planets should examine more than one sample, which means that either several samples of different materials have to be brought to the instrument or the instrument itself has to be moved to the different samples, like that a wide spectrum of results gives a better view of the explored target. The current trends for the infrastructure in space technology can be formulated with:

- An increasing reach and mobility of space systems (meters to km)
- An increasing autonomy requirements
- Mobility operations requirements and analytical capabilities

To meet these mobility requirements the trends for the instrumentation are as follows:

- A general miniaturisation (decreasing mass, size, resource needs)
- The change from point measurements to repetitive and systematic analysis
- Criticality of measurement time
- The development from single instruments to combined instruments or instrument facilities (intra–P/L integration)

A variety of rover systems have been developed in the previous years in order to accomplish all these needs for the exploration of target surfaces. Theses systems differ in their range, complexity, their demands on control, power needs etc. Thus a rover system can be adjusted to the expected mission.

H.5 Small Systems

H.5.1 Nanokhod

The Nanokhod is a small rover system which is fulfils the demand for mobility specifically in the vicinity of a lander unit.

The microrover was developed and designed by the company *von Hoerner & Sulger* (vH&S), from an initial Russian concept. Within the history of the Nanokhod development the rover has been designed for the environment conditions of the planets Mars as well as Mercury (both, hot and cold side) [4, 5].

Figure H.2: The microrover Nanokhod while positioning the Scientific Payload Instruments

The microrover has a volume of $160 \times 65 \times 250\,mm^3$ and weighs 3.2 kg including a payload mass of 1kg. The payload as implemented for the Mercury design [3] consists of two spectrometers and one micro imager a highly integrated form so that it fits into the available mass and volume. The Nanokhod is a tethered system, that uses the lander as its power supply (with a peak power is 5 W) and its data relay to Earth. The tether connection allows a mobility range of 50 m with an expected radius of 5m around the lander, while the maximum speed of the Rover is up to 5 m/h.

The Nanokhod is a small mobile scientific platform, which is able to carry a package of scientific instruments to the different points of interest around the lander, and thus allows in–situ exploration of rocks and small craters in the near vicinity of the landing point.

H.5.2 Lunar Mission BW1

This mission [6] is the forth project of the Stuttgart Small Satellite Programme of the Institute of Space Systems of the University of Stuttgart. The all electrical small lunar satellite with a size of 1m cube, a mass of around 200 kg and 1 kW electrical power is planned to be launched after 2008.

The satellite should be inserted into a GTO to use its two different electric propulsion systems (thermal arcjet, magneto–plasmadynamical thruster) to spiral out to lunar vicinity and reach a high–inclined low lunar orbit. After a scientific operation phase of six month performing different remote sensing experiments the satellite should impact on the lunar surface.

Even if the payload is not yet decided multispectral camera systems in the visible and infrared range as well as radar sounding using the communication system are conceivable.

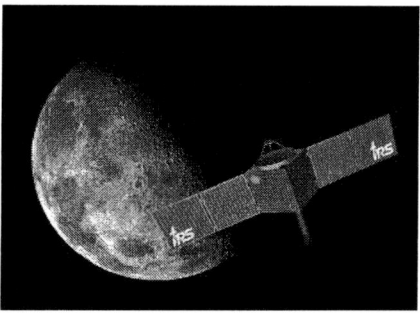

Figure H.3: Lunar Mission BW1

H.6 Feasibility Study of the Micro Rover Integration

Two feasibility studies [1, 2] were carried out in order to check the opportunities of using the small satellite Lunar Mission BW1 to carry a lander with an adapted Nanokhod as a payload to the surface of the Moon. This would be done during the decent phase of the impact experiment at the end of the mission. As the satellite Lunar Mission BW1 itself is still in the development phase the studies for a feasible payload had to be held into a wider frame, so in the end a payload can be designed for each available mass and volume share on the satellite. This gave a wide range of results and prepared the payload for several different requirements that could become possible.

The six different concepts varies in different possibilities of locomotion and payloads, see table H.1.

Concept	Payload	Terrainability of the rover	Connection to lander/ Communication
A	Camera	Driving straight	Sender option
B	Camera	Driving straight, parallel movement of PLC	Sender option
C(T)/ C(R)	Camera	Disconnected LU movement, parallel movement of PLC	Sending/Tether connection, Sender and Receiver option
D	Camera, APXS	Driving straight	Tether connection, Sender and Receiver option
E	Camera, APXS	Disconnected LU movement, parallel movement and rotation of PLC	Tether connection, Sender and Receiver option
F	Camera, APXS, MIMOS	Disconnected LU movement, parallel movement and rotation of PLC	Tether connection, Sender and Receiver option

Table H.1: Overview of the considered rover concepts

H.6.1 Payload

Because of volume and mass restrictions only a small number of possible payloads was part of the study. To fulfil the primary mission objectives a small CCD/CMOS matrix camera with a mass of 100 g is part of all rover concepts. Additionally a Alpha Particle X–Ray Spectrometer (APXS) with a mass of 370 g for chemical analysis and a Miniaturised Moessbauer Spectrometer (MIMOS) for mineralogical analysis with a mass of 400 g can be part of the scientific payload.

H.6.2 Communication

For the link between the Nanokhod and the ground station two different concepts of communication were considered: one way communication for the simple rover concepts and two way communication for the more complex rover concepts. Link budgets taking different frequency bands starting from VHF up to Ka–band and various antenna types into account were compared.

The suggested system uses S–band (2 GHz) with an elementary dipole at the rover and 3 W transmitting power (10 W input power). This results in approx. 2 kbps data rate with a 27 m parabolic antenna on Earth instead of using the institute's own ground station (2.5–3 m parabolic antennas). Because of the short mission duration renting a larger dish seems to be feasible.

Usage of a parabolic reflector on the rover or lander would increase the transmission rate but a complex alignment mechanism would be necessary. Higher data rates at lower frequencies are available by using other lunar orbiter missions as relay stations.

H.6.3 Power Supply

Depending of the rover concept the power consumption is assumed with 34–48 Wh/day. Compared to the limited locomotion distance of the tethered system (50 m) the "free" Nanokhod is able to drive up to approx. 100 m/day. Especially the number of scientific instruments drives the power consumption.

Because of its high specific energy–mass ration Li–Ion batteries were chosen for the mission. 2–3 cells provide approx. 18 Wh EOL power – taken into account the transfer time of two years or more during the cruise flight on–board of the Lunar Mission BW1. The batteries have to be recharged during the mission by a solar array of less than 150 cm^2 providing max. 3 W using multi–junction GaAs solar cells – the here assumed values are for the situation seven days after sun rise at 50 deg latitude.

H.6.4 Thermal Control

The Moon is not a spacecraft friendly environment even in thermal affairs. Two different cases were defined and inquired: hot case (full operating rover, max. sun elevation on the 7th day after sun rise) and cold case (rover in standby mode, low elevation of the sun 14 hrs after sun rise).

Taking into account the temperature limits of subsystems and the components as well as the temperature range at different latitudes the rover demands a radiator on top but can be heated with the internal subsystems during operations.

H.6.5 Lander

The micro lander should protect the rover during separation from the satellite, descent and landing on the lunar surface. The landing would comprise the following phases: the cylindrical spacecraft would be ejected from the satellite before its impact on the Moon. A solid rocket motor is ignited for approx. 10 s to reduce the velocity before separated from the lander. To land softly an inflatable structure ("airbags") is used.

The minimum mass of the lander without any carried rover is approx. 13 kg (incl. up to 50% margin). Especially the thrust error of the solid rocket motor is a major driver of the mass budget.

H.7 Results

The analysis shows that an adapted Nanokhod can be landed and operated on the lunar surface at various latitudes for up to nearly one lunar day. Depending on the level of complexity of the chosen concept and the number of instruments the rover mass is between 2.5 and 5.5 kg (incl. 20% margin). The total lander mass varies between 18.5 and 26.5 kg (incl. up to 50% margin)

H.8 Outlook

Further detailed joined vh&S/IRS–studies to design and build prototypes of one of the lander and rover concepts are planned for the near future. Especially the suggested landing system should be tested using a lander prototype model.

H.9 References

1. Nagy, M.: Mission Analysis and Conceptual Design of a Lunar Micro Lander Mission, Research Paper, IRS–05–S–09, Institute of Space Systems, Universität Stuttgart, 2005

2. Graf, T.: Conceptual Design of a Lunar Micro Lander and Corresponding Feasibility Study, Research Paper, IRS–05–S–29, Institute of Space Systems, Universität Stuttgart, 2005

3. Schiele, A., Romstedt, J., Lee, C., Henkel, H., Klinkner, S., Rieder, R., Gellert, R., Klingelhöfer, G., Bernhardt, B., Michaelis, H.: The new NANOKHOD: Engineering model for extreme cold environments, 8th International Symposium on Artificial Intelligence, Robotics and Automation in Space, September 5 – 8, 2005, Munich, 2005

4. Micro–Robots for Scientific Application,. Summary Report, ESTEC Contract No.
12052/96/NL/JG(SC), von Hoerner & Sulger GmbH, Schwetzingen, Germany, 1999

5. Robotic Technology for Planetary Exploration, Executive Summary, ESTEC Contract No.: CCN2 to 12052/96/NL/JG(SC), von Hoerner & Sulger GmbH, Schwetzingen, Germany, 2003

6. Laufer, R., Roeser, H.–P. & The Lunar Mission BW1 Project Team: Small Satellites for Future Cislunar and Lunar Infrastructure – 'Lunar Mission BW1' of the University of Stuttgart, 9th ISU Annual International Symposium, November 30 – December 3, 2004, Strasbourg, 2004

Curriculum Vitae

Sabine Klinkner

Date and place of birth   November 24, 1975 in Greenwich, CT., USA

Education/Research

05/1995	Hegel– Gymnasium, Stuttgart: Abitur (high school graduation exam)
09/1996–02/2001	Stuttgart University: Undergraduate and graduate studies in Aerospace Engineering, incl. Major in Thermodynamics and Space Systems
03/2001–10/2001	ITLR (Institute of Thermodynamics) of Stuttgart University: Research project
01/2002–08/2002	Centrospazio (connected to Pisa University), Italy: Diploma Thesis;
09/2002	Graduation as "Diplom–Ingenieurin" in Aerospace Engineering
Since 10/2004	Scientific staff at the Institute of Space Systems, Universität Stuttgart, Research field "Small Satellite Programme" Accomplishment of the doctoral thesis with the title: *Concepts and development of highly integrated payloads for space missions on the example of the microrover NANOKHOD*

Professional Experience

11/1995–01/1996	Working experience, Innenarchitekturbüro SPEK–Design, Stuttgart
02/1996–03/1996	Practical taining, Metallverarbeitungsbetrieb Wittenbecher GmbH, Berlin, Germany, Introduction in the basics of metal processing
04/1996	Practical taining, Metallgiesserei Weigel, Stuttgart, Germany, Introduction in the basics of foundry
07/1996	Practical taining, Metallwarenfabrikationsbetrieb Werner Hermann, Stuttgart, Germany, Introduction in the basics of welding

10/1998–11/1998	Quality Management/Engineering project, Metallver-arbeitungsbetrieb Wittenbecher GmbH, Berlin, Germany
12/1998	Quality Management/Engineering project, Phillips Admeco, Eindhoven, Holland
01/1999–03/1999	Light plane floats development, Australian Light Wing, Ballina, NSW, Australia
Since 10/2002	von Hoerner & Sulger GmbH, Schwetzingen, Germany, Sub–project management, robotic system developments, mechanical design and structural and thermal CAE on various space projects in the fields of rover technology, space instrumentation and camera systems for space applications

Received Awards

2007	Amelia Earhart Fellowship
01/2005–12/2007	Irene Rosenberg grant for doctoral project, Ministry of Science, Research and the Arts, BW
2002	Erasmus–funding, part of Sokrates programme, funded by the EU
2001	Herman Reissner grant, Herrmann Reissner Foundation – department of Aerospace Engineering, Universität Stuttgart
1994	"Jugend Forscht" (Youth research programme), exceptional price in the physics sector